THE BEER BOOK

Anderson & Co., Home Brewed Ale,
Albany, N.Y., 1862-1872

In today's "plastic and aluminum world" it is hard to remember when beer barrels were made of wood. The barrel shown here is from the Kips Bay Brewery, a New York City brewery that produced only draft beer throughout its entire history (1894–1948). Its plant at 1st Avenue and 38th Street is still standing and now serves as the world headquarters for CARE.

An Illustrated Guide

THE BEER BOOK

to American Breweriana

by Will Anderson

THE PYNE PRESS • PRINCETON

First edition

Library of Congress Catalog Card Number 73-79519

SBN 87861-057-X

*Printed in the United States
of America*

Designed by Gwendolyn O. England

Printed at Pearl Pressman-Liberty, Philadelphia

Contents

Preface

The purpose of *The Beer Book* is extremely basic—to entertain and to inform. Hopefully, it will make more people realize that collecting breweriana is a great pastime. It is a hobby that literally can make history come alive—and alive with beauty and color.

If, after reading the next 200 pages, you come away with a desire to brighten your home and your life with a collection of brewery trays or lithographs or bottles or whatall, then this book will have been successful. And, if you are further left with the desire to delve into the history behind the items you collect, then our efforts will have been doubly successful.

The Beer Book is divided into five sections, the first of which offers a definition of breweriana, explains the factors that make it such a rewarding and entertaining collecting field, and tells where to begin your collecting search. The second and largest section presents, one by one, the myriad of breweriana collectibles. Twenty-one different types are included, and each is fully illustrated.

Specific American breweries is what the third section is all about. Thirty-eight have been chosen as representative of the last century in U. S. brewing. For each you'll find an informative and readable capsule history, plus some fine photographs of advertising and packaging items used by the companies through the years.

Great brewing cities is the theme of the fourth section. Each major center is included in picture and story. And, finally, in the fifth section, we present a short but fascinating survey of American beer fact and fancy.

• • •

Writing a book—especially when you love the subject—is fun. But it is also hard work, and something that is impossible to do alone.

To my wife, Sonja, I give the warmest and most loving "thank you" imaginable. Without her typing, bottle painting and, above all, constant encouragement, *The Beer Book* would never have become a reality.

I'm also deeply grateful to Steve Seidel of Utica, N. Y. Steve, a good collecting friend and President of the Eastern Coast Breweriana Association, heard about the book and offered to lend me any or all of his very sizeable collection. As a result, a number of fine trays and other items are included in *The Beer Book* that otherwise would have been missing.

Joel Kopp of America Hurrah Antiques was also generous enough to lend me several items from his shop in New York City.

For the use of their library facilities, and especially for all the cooperation and information they provided, I wish to thank Miss Pat Jones and Mr. Philip Katz of the United States Brewers Association.

Considerable knowledge has been accumulated by a number of leading breweriana collectors. For invaluable help in their fields of expertise, I am indebted to:

Leon Beebe, Mt. Airy, Md., glasses and mugs; Thomas Dallman, Wauwatosa, Wis., general items; Bob and Jean Fugina, Sacramento, Calif., miniature beers and Buffalo Brewing Co.; Robert Gottschalk, Rochester, N. Y., Rochester breweries; Louis Greco, Bronx, N. Y., foam scrapers; Herbert Haydock, Wisconsin Rapids, Wis., coasters; John Murray, Hinsdale, Ill., glasses and mugs and general items; Jack Muzio, Santa Rosa, Calif., trays; Robert Myers, Oakland, Calif., cans; Donald Rathbun, Madison, Wis., picnic bottles; Joseph Veselsky, Hicksville, N. Y., labels.

Special thanks are due to Jack Muzio and Bob Myers for complete use of the fine material they have published.

I wish also to thank the many breweries, historical associations, libraries, chambers of commerce, and other organizations that were good enough to answer my pleas for help:

Albany Institute of History and Art, Albany, N. Y.; American Can Co., Greenwich, Conn.; Anheuser-Busch, Inc., St. Louis, Mo.; Blitz-Weinhard Co., Portland, Ore.; Chattanooga Public Library, Chattanooga, Tenn.; Cincinnati Historical Society, Cincinnati, Ohio; Corning Museum of Glass, Corning, N. Y.; Douglas County Historical Society, Omaha, Neb.; Falstaff International Museum of Brewing, St. Louis, Mo.; Genesee Brewing Co., Rochester, N. Y.; Theodore Hamm Co., St. Paul, Minn.; G. Heileman Brewing Co., La Crosse, Wis.; Historical Society of Berks County, Reading, Penn.; Historical Society of Pennsylvania, Philadelphia, Penn.; Jones Brewing Co., Smithton, Penn.; The Lion, Inc., Wilkes-Barre, Penn.; Maryland Historical Society, Baltimore, Md.; Mechanicville Area Chamber of Commerce, Mechanicville, N. Y.; Memphis Public Library and Information Center, Memphis, Tenn.; Miller Brewing Co., Milwaukee, Wis.; Milwaukee County Historical Society, Milwaukee, Wis.; Missouri Historical Society, St. Louis, Mo.; New Bedford Area Chamber of Commerce, New Bedford, Mass.; New York State Alcoholic Beverage Control, New York, N. Y.; Old Crown Brewing Co., Fort Wayne, Ind.; Onondaga Historical Association, Syracuse, N. Y.; Olympia Brewing Co., Olympia, Wash.; Pabst Brewing Co., Milwaukee, Wis.; Pearl Brewing Co., San Antonio, Tex.; Reading Brewing Co., Reading, Penn.; *Register-Star,* Hudson, N. Y.; Rensselaer County Historical Association, Troy, N. Y.; Rheingold Breweries, Brooklyn, N. Y.; Sandusky Area Chamber of Commerce, Sandusky, Ohio; Jos. Schlitz Brewing Co., Milwaukee, Wis., and Brooklyn, N. Y.; C. Schmidt & Sons, Norristown, Penn.; Seattle Historical Society, Seattle, Wash.; Spoetzl Brewery, Shiner, Tex.; State Historical Society of Colorado, Denver, Colo.; Stegmaier Brewing Co., Wilkes-Barre, Penn.; Sterling Brewers, Evansville, Ind.; Stroh Brewery Co., Detroit, Mich.; Walter Brewing Co., Pueblo, Colo.; West Bend Lithia Co., West Bend, Wis.; West End Brewing Co., Utica, N. Y.; Whiskey Gulch Gang, Inc., Canyon City, Ore.; Geo. Wiedemann Brewing Co., Newport, Ky.; D. G. Yuengling & Son, Pottsville, Penn.

Introduction

"Breweriana" is a word that was coined several years ago to signify any and all items relating to beer and brewery advertising and packaging. More specifically, this includes trays, signs, cans, advertisements, bottles, labels, glasses, coasters, taps, openers, bottle tops, mugs—an almost endless array of objects.

People collect things for many different reasons. Generally, however, two motives stand out from all others—fun and profit. Breweriana, one of the fastest growing American hobbies, fulfills both of these goals very nicely.

The American brewing industry has had a fascinating history. Going back to the early days of the country, almost every household was a "brewery," in the same sense that each was also a "bakery" and a "farm." Americans were largely self-sufficient and malt beverages were "home-brewed." The beginnings of the Industrial Revolution in the first half of the nineteenth century, however, changed all this, and by the middle 1800's most malt beverages were commercially brewed. In fact, by 1850 there were 431 commercial breweries in the United States, as compared with only 129 in 1810. In 1860 there was 1,269 and by 1873 this figure had risen to a remarkable all-time high of 4,131. Those were the days when almost every town of any consequence had its own brewery and its own beer.

What could be more interesting or entertaining than the search for these local items? Almost regardless of where one lives, it is possible to ferret out facts about, and items from, local breweries—whether "local" means a particular town, a particular state, or even a particular group of states.

Equally fascinating, however, as a local collection is the reverse—a collection representing brands and breweries from all over the country, or even the world. Beer, unlike many mass produced items, has had few national brands. This, of course, was especially true before the great rise in packaged beer sales following the repeal of Prohibition. But it is true even today. Such extremely well-known brands as Coors, Lucky Lager, Pearl and Schaefer are not distributed nationally. And just try to find Anchor Steam outside of the immediate San Francisco area or Horseshoe Curve in any other place but Altoona, Penn. As a result, a business or pleasure trip to almost any locale is a lot more interesting, whether you are a bottle or tray collector stopping at antique shops and flea markets, or a can collector stopping at grocery and package stores.

Regardless of whether a collection is local, national or worldwide, half the fun can be found in historical detective work: what breweries were in operation and where; what brands were produced; how long they were in operation, etc. This is an aspect of breweriana that can greatly enrich collecting, and one that should not be overlooked.

While all of the above "for the fun of it" reasons for collecting breweriana have considerable validity, there is one additional reason that towers above all others—its decorative value. Beer advertising and packaging has always been extremely colorful and ornate. The hundreds of pictures in *The Beer Book* are testimony that a thing of beauty is a joy forever.

In addition to the fun of collecting, breweriana has been, and should continue to be, an extremely good investment. The prices on antiques and collectibles in general have risen considerably over the past decade, and this seems to be especially true of the newer types of Americana—items from the turn-of-the-century up through the 1950's.

Most breweriana fits into this last category, and the prices for trays, signs and other advertising objects have risen astronomically in some areas of the country during the last five to ten years. This may have been caused in part by the greatly increased interest in old advertising items used to decorate home bars, dens and playrooms. The use of these items in Gay 90's restaurants and taprooms has also increased demand. However, the basic reason for the rise in value is, naturally, the great increase in private collecting. As leisure time becomes more plentiful, more people channel their energy into hobbies of all kinds.

In addition, the high brewery mortality rate has also had a very definite effect on prices. An empty can of a brand made today generally sells for 25¢ and a tray for $1 to $2.50 or so. Just let that brewery go out of business, however, and the can will rise in price to 50¢ or $1 and trays will reach the $3 to $6 range. This increase is due not only to the "romanticism" attached to out-of-business companies, but also because of the realization that these items, no matter how plentiful up until the present, will never be manufactured again. A glance at the following list will indicate the strong impact of brewery closings:

year	U. S. brewery plants in operation
1860	1269
1870	1972
1880	2272
1890	1928
1900	1758
1910	1498
1916	1313
1935	750
1940	611
1945	468
1950	407
1955	292
1960	229
1965	197
1970	142
1972	136
1975	?
1980	??

It should be pointed out that this chart is somewhat misleading in that it shows brewing plants and not brewing companies. In 1972, for example, ten companies operated 53 of the total 136 plants. The big keep getting bigger, while the small and medium-sized breweries are finding it harder and harder to compete.

For a free list of brewery plants in operation for the present year, write to the Bureau of Alcohol, Tobacco and Firearms, Department of the Treasury, Washington, D. C. Ask for publication #659, "Breweries Authorized to Operate." A directory containing information on breweries world-wide can be purchased from *Modern Brewery Age,* 80 Lincoln Ave., Stamford, Conn., 06904. This directory, known as the *Modern Brewery Age Blue Book,* is published annually in the Spring.

Searching for breweriana objects is more than half the fun. The collector can look to three major sources for items.

First and foremost for all categories, except cans and labels, are flea markets, antique shows and shops and garage/tag sales. Flea markets especially should not be missed. It's amazing what great items often turn up, and sometimes at bargain prices. Antique shops are generally not as rewarding as the markets, but if you cover enough of the shops you'll undoubtedly find a few that specialize in "advertiques." Cultivate these speciality outlets, and you'll find that you have a standing source of ever-changing material. Many antique dealers keep a "want" list and, if asked, will be glad to notify you if they run across any interesting beer items.

In the past few years it has been possible to buy from (or at least view the wares of) many such "advertique" dealers under one roof at the antique advertising shows that have been held in various parts of the country. One such show that we recently attended, for example, took place at the fairgrounds in Gaithersburg, Md. There were approximately 100 dealers, all loaded with trays, signs, posters and all kinds of other outstanding advertising items. In fact, there were more breweriana collectibles on hand than otherwise could have been seen in a year. The only problem with these shows is that the prices are sometimes quite a bit above what you might otherwise pay.

The best way to keep posted on what flea markets and shows are being held in your area is to read the local newspaper and, more importantly, to subscribe to one or more of the special collecting periodicals.

As is true with so many other types of merchandise sold in the United States, there is very definitely a mail order market. In fact, an individual could build a fairly good collection of trays, bottles, cans, or most any other breweriana item without ever leaving his home. How? Simply by following the items listed for sale in the numerous collecting newspapers or magazines. These are, generally speaking, tabloid size newspapers containing page after page of classified ads under such categories as "Bottles," "Books," "Political Items," "Advertising Items" and, of course, "Miscellaneous." Among the collectors' papers that we've found most productive are:

American Collector, 13920 Mt. McClellan Blvd., Reno, Nev. 89506
Antique Trader, Box 1050, Dubuque, Iowa 52001
The Antiquity, Hope, N. J. 07844
Collector's News, Box 156, Grundy Center, Iowa 50638
Collector's Weekly, Drawer C, Kermit, Texas 79745
Joel Sater's Antique News, Box B, Marietta, Penn. 17547

Subscriptions to these periodicals are reasonable—and invaluable to any collector, whether a beginner or a real veteran.

The third major source of breweriana is, naturally, that of other collectors. The best way to meet fellow collectors and to form a valuable trading link is to join one or more of the several breweriana clubs that have been formed in the past several years:

Eastern Coast Breweriana Association. Despite its name, this is a national organization concerned with all collectibles. Trading session meetings (thus far, always in the East) are held four or five times a year at different members' homes. ECBA inquiries should be directed to Steve Seidel, President, ECBA, 135 Ridge Road, Utica, New York 13501.

National Association of Breweriana Advertising. If you collect (or think you might like to collect) trays, signs or other advertising material, then you should very definitely join NABA. A weekend-long convention is held annually and includes one or more brewery tours, knowledgeable speakers, and a buy/sell/trade session. Inquiries should be directed to John Murray, President, NABA, 475 Old Surrey Road, Hinsdale, Ill. 60521.

Beer Can Collectors of America. If beer cans are your main area of interest, this is *the* group. The club sponsors a convention for one weekend a year and a buy/sell/swap session that features more cans than you'd normally see in a lifetime. For information, contact Beer Can Collectors of America, P. O. Box 9104, St. Louis, Mo. 63117.

How much should you pay for various items? This is a difficult matter to discuss. Trying to price breweriana is virtually impossible because of the great price variations that occur not only in different parts of the country, but also in different parts of the same flea market. The same tray, sign or bottle can sometimes be found priced three or four different ways at a good-sized antique show or flea market. A tray may be $15 at one booth, while four booths away it has a $25 label, and—3,000 miles away—it's $40 to $50. As a result, there can be no standard set of prices and to even attempt a compilation is folly. There are, however, some *very general* outlines for three of the major collecting categories:

Bottles. This is the one breweriana category, in most of the East and Middle West, at least, that has not risen greatly in price. Good embossed blob and older crown bottles are still available in the $2 to $5 range. On the West Coast, prices for good beers are considerably higher, ranging from $8 to $15 for more common bottles.

Trays. These comprise the hardest category to price. Generally speaking, good trays have gone through an unbelievable inflationary period in the past five years. Gone forever are the days when you could discover a stack of good $2 to $3 trays half-hidden under a flea market dealer's display table. Now these prices will afford you a tray containing more rust than metal or one so new that the paint has barely dried. Therefore, our only guideline is—be prepared to pay plenty for older trays in desirable condition. The starting point is $10 to $15, and the going price for fine examples is up around $50 to $100. Once again, West Coast (especially California) trays command a much higher price. Basic supply (comparatively few West Coast breweries) and demand (California may have more serious tray collectors than any other two states put together) dictate this substantial price differential.

Cans. As noted earlier, these items are still low in price, although they are suffering from the same inflationary trends as older items. Current brands usually bring 25¢ while old designs of still-in-existence brands sell for 50¢ to $1. Pre-1950 flat-tops in very good condition will generally run several dollars while spout tops, the goal of every collector, are now up to $4 to $7 each.

Regardless of category, condition is all important. No collector wants to display a rusty tray, pitted can or chipped bottle or glass. At flea markets and dealers' shops, of course, the condition of the piece is easily gauged—"what you see is what you get!" In buying through the mail, however, it is not quite that simple. You should be sure that condition is clearly stated by the seller.

Collectors of breweriana are well advised to familiarize themselves with the different major types of malt beverages, which vary greatly in popularity and availability:

Ale. A fermented malt beverage which is flavored with hops, generally heavier bodied, and containing a higher alcoholic content than beer. Ales can vary greatly in color and bitterness/sweetness. Albany, N. Y., was probably America's greatest ale-producing city and many famed brewers either were located or started there, including Peter Ballantine. Though best remembered for the Newark, N. J., brewery he founded in 1840, Ballantine was a member of the Albany ale brewing fraternity from 1833 to 1839.

Porter. A dark brown, heavy malt liquor resembling ale. It acquired its unusual name because of its early popularity with English porters. Philadelphia was once famous for this brew.

Stout. A malt beverage similar to porter, but much darker in color and with a strong malt flavor and sweet taste.

Lager Beer. A bottom-fermented beverage as opposed to ale which is top-fermented. Lager is a German word meaning storehouse. Lager beer has to be stored in a cool place to age properly, while ale ferments much more rapidly and requires a warmer temperature.

Bock Beer. A heavier, darker and richer type of beer than lager. Bock is generally aged over the Winter months and sold in the Spring. Legend has it that this special brew was born in Einbeck, Germany, in the year 1250. Through the years Einbeck was corrupted to Einbock. In any case, bock is the German word for the male goat, which explains why virtually all bock beer ads and labels feature a well-horned goat.

Pilsener Beer. A light, pale lager beer, named after Pilsen, Czechoslovakia, where it was originally brewed. Most American beers are of this type.

Weiss (white) Beer. A pale, effervescent beer, usually brewed from wheat that goes through its second fermentation in the bottle.

Malt Liquor. Usually darker in color and somewhat more bitter than traditional American lager. The main difference, however, is that it has a higher alcoholic content. Lately, in an attempt to counter the great increase in popularity of sweet wines, several U. S. breweries have been marketing beer and wine or beer and soda combinations as malt liquor.

Lager beer, which accounts for a good 90% of the malt beverages sold in the United States today, was not introduced into this country until 1840. Up to that time production had been limited to malt beverages in the English tradition—ale, porter and stout. The brewer given credit for introducing lager brewing to America is John Wagner, who had a small brewery in the rear of his house on St. John and Poplar Streets, Philadelphia. Wagner brought the first lager beer yeast with him from Bavaria, where he had been a brewmaster.

This new German-type beer took awhile to catch on, but within a generation it was outselling all its competitors. Popularity was greatly advanced by the vast number of German immigrants that settled in the United States during the mid-nineteenth century. German communities were especially strong in Philadelphia, Brooklyn, Milwaukee, Cincinnati, St. Louis and Chicago, and it was naturally in these cities that lager beer production and consumption were heaviest.

There is one type of malt beverage that deserves to be singled out because of its importance to bottle collectors—Weiss beer, a variety that has almost totally disappeared from the U. S. market.

Back in the second half of the nineteenth century, Weiss beer was produced by many U. S. breweries. Generally these breweries specialized in Weiss, and, most often, they were quite small with respect to output. It's believed they also bottled virtually all of this output. As a result, a goodly percentage of early beer bottles are Weiss beers. In 1884, for example, of the 35 breweries in Brooklyn, only nine had bottling works and eight of these nine were Weiss breweries.

It appears that no American brewery has produced Weiss since Prohibition, and certainly there is none that produces it today. Imported Berliner Weiss is available, however. This is a special type that is generally mixed with raspberry syrup and drunk from a large goblet. It tastes more like

All 28 of these bottles are embossed with either Weiss Beer or Bier, usually on the reverse. Weiss beers are generally of the type known as "squat blobs." St. Louis Weiss beer bottles, however, had their own characteristic shape, as shown here by the two taller bottles in the second row. Regardless of their shape, virtually all of the bottles pictured pre-date 1900, some by quite a few years. George Esselborn, for example, changed his New York location in 1880 or 1881 from 50th Street & Broadway (2nd bottle from left, 3rd row) to 613/615 W. 47th Street (extreme left bottle, 3rd row). In addition, the company is listed as George Esselborn's *Sons* in 1890. Both bottles, therefore, must predate 1890, and the one with the Broadway address goes back to the 1870's.

wine than beer and, in fact, was given the nickname Le Champagne du Nord (The Champagne of the North) by Napoleon on one of his German campaigns.

The Beer Book is not meant to be a history of U. S. brewing, although we have attempted to include considerable historical fact and information throughout. There is one period, however, in the industry's history that does deserve special attention because it serves as the dividing line between old and new for virtually all breweriana collectibles. This period is Prohibition.

Temperance forces were active in the United States since the early 1800's. The first organization of any consequence was the American Society for the Promotion of Temperance (usually called the American Temperance Society), founded in Boston by the descendants of beer-drinking Puritans in February of 1826. In the next 90 years many other groups, some for religious and some for purely moral reasons, sprang up to fanatically fight the evils of alcohol. The two most effective opponents have been the Women's Christian Temperance Union (W.C.T.U.), founded in Ohio in 1874 and now headquartered in Evanston, Ill., and the Anti-Saloon League, organized on a national scale in 1895.

While temperance was their name, total abstinence was generally the goal of these organizations. And they were tough to fight. Godliness, Motherhood, Happy Home Life—all, and more, were invoked in the name of temperance.

TRUE TEMPERANCE.

STATISTICS prove that drunkenness has decreased to a remarkable degree in the last thirty years, which is absolutely attributable to the use of Beer (containing 3½ to 4 per cent. alcohol) instead of the more ardent spirits.

Beer, therefore, is recognized by Physicians and Scientists as one of the principal promoters of TRUE TEMPERANCE.

The average drinking man (or woman) had no protection against such fervor. And the brewers, instead of combining with the distillers to form a united front, often appeared to oppose hard liquor themselves. Hoping that beer would be exempted from any possible prohibition legislation, they concentrated on a "Beer is the pure temperance drink" theme.

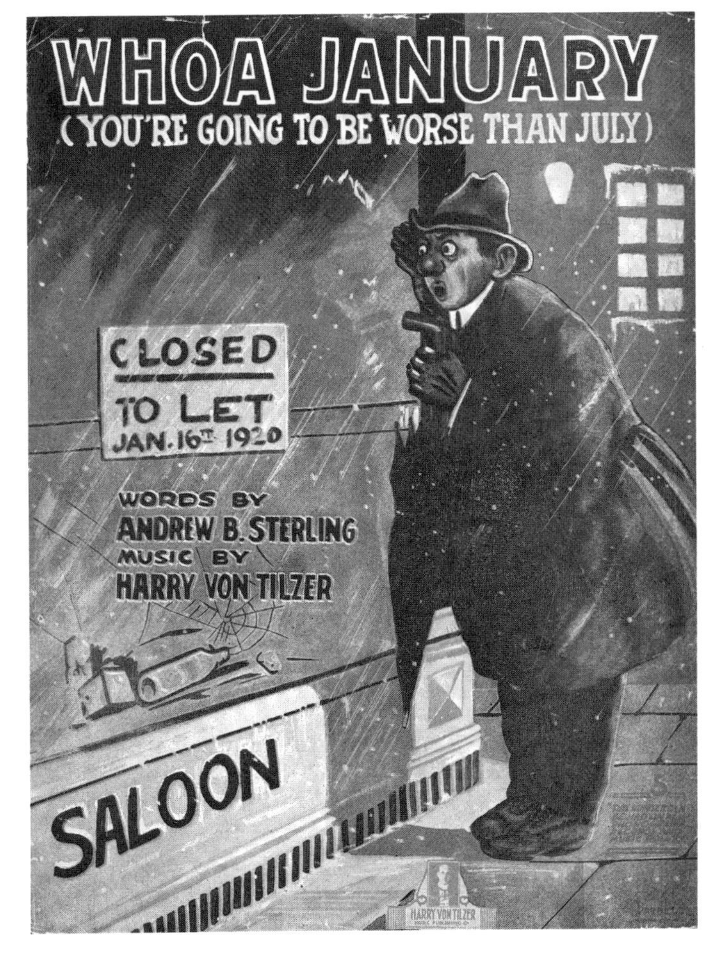

When it came, Prohibition was complete. The 18th Amendment to the Constitution forbade "the manufacture, sale, or transportation of intoxicating liquors within, the importation thereof into, or the exportation thereof from the United States" and all its territories. Any beverage, including beer, that contained more than one-half of one percent alcohol by volume was classified as an "intoxicating liquor." The 18th Amendment was ratified January 16, 1919, and became effective one year from that date, on January 16, 1920.

Their livelihood legislated away, most brewers went out of business while the rest of the nation roared through the 1920's. Those that did manage to stay open turned to the manufacture of ice cream, spaghetti, chocolates, cured ham, etc., etc.—anything to remain solvent. The most successful products were soft drinks and, of course, near beer containing less than one-half of one per cent of alcohol. Syracuse's George Zett Brewery, for example, manufactured Par-Ex near beer and root beer, and Zett's "Snappy Beverages" (20 flavors!) throughout the 1920's.

"All the News That's Fit to Print."

The New York Times.

LATE CITY EDITION

VOL. LXXXII....No. 27,467. NEW YORK, FRIDAY, APRIL 7, 1933. TWO CENTS

BEER FLOWS IN 19 STATES AT MIDNIGHT AS CITY AWAITS LEGAL BREW TODAY; 3.2 ERA OPENS HERE WITH FEW REVELS

ROOSEVELT TO ASK 4 NATIONS TO TALKS; M'DONALD ACCEPTS

HERRIOT VISIT SUGGESTED

Delegates From Italy, Germany and Japan to Be Welcome Later.

'REALISTIC ACTION' SOUGHT

President Tells MacDonald He Wants 'Practical' Arms and Economic Talks.

WHITE HOUSE OPEN TO HIM

Prime Minister to Stay There—Envoys Flock to State Department With Inquiries.

By ARTHUR KROCK.

Student Strike Over Job Law Makes Sorbonne Halt Classes

SAW AKRON GIRDERS BUCKLE IN STORM

Deal, Quoted in Wiley Report to Swanson, Said Two Bent As Gusts Hit Ship.

HOUSE INQUIRY SPEEDED

Vinson Suddenly Orders It to Begin Today—Roosevelt Greets 3 Survivors.

Senate Votes 30-Hour Work Week, 53-30; Robinson 36-Hour Amendment Is Beaten

Continued on Page Seven.

NAZIS SEIZE POWER TO RULE BUSINESS; OUR FIRMS ALARMED

Entire Board of Federation of German Industries Resigns, Commissioners Stepping In.

OLD TESTAMENT IS BANNED

Hitlerites Demand German Sagas Replace First Part of Bible in Churches.

By FREDERICK T. BIRCHALL.

BUREAUS WIPED OUT, 'DEADWOOD' CUT OFF BY ROOSEVELT AXE

'Reorganization' Now Becomes 'Elimination' Under Orders to Reach Economy Goal.

SCIENCE SERVICES FALL

Agriculture, Interior, Commerce and Postoffice Departments Hardest Hit.

MANY CITIES CELEBRATE

Cheering Crowds Hail Beer Trucks Distributing New Brew.

CAPITAL LEADS THE WAY

Philadelphia, St. Louis, Baltimore, Milwaukee and San Francisco Hold Gala Night.

JOLLITY REIGNS SUPREME

Absence of Rowdyism Marked as Thousands Flock to the Hotels and Restaurants.

Roosevelt Gets First Cases of Capital's 3.2 Beer; Gov. Lehman Will Serve the Brew at Albany

ALBANY DEADLOCK CONTINUES ON BEER

Lehman Sticks to His Control Plan as Legislature Nears Adjournment or Recess.

SUNDAY SET AS TIME LIMIT

Governor Signs Bill for $1-a-Barrel Tax on Brew and Measure for Repeal Convention.

BEER PERMIT RUSH LASTS TO MIDNIGHT

City Offices Stay Open After Snarl Ties Up Licenses Until the Afternoon.

CROWDS STAND IN RAIN

40,000 to 60,000 Expected to Sell Brew Here—Blanks of Fire Department Used.

BROADWAY DISAPPOINTED

Hopes of Carnival on New Brew Fade as Only Bootleg Is Served.

CROWDS ARE APATHETIC

110 Extra Policemen Guard White Light Area, but No Need for Them Arises.

BREWERIES DRAW MANY

Throng in Brooklyn Tries to Seize Bottles on Truck, Ready for Early Deliveries Today.

Prohibition ended—for beer and wine—on April 7, 1933. Eight months later, on December 5, 1933, the 21st Amendment repealed the 18th entirely. Drinking was legal again and the brewers returned to their kettles, undoubtedly with a smile on their face and a sparkling lager in their hand. *Reprinted by permission of The New York Times*

THE COLLECTIBLES

As opposed to pre-Prohibition covered cases, this is a fine example of a 1930's "topless" case, from the Welz & Zerweck Brewing Co., Brooklyn, N.Y.

Bottles

Although bottled beer is now taken for granted, one hundred years ago such was far from the case. Beer was generally sold by the keg, and what little was bottled had to be consumed within a few days before it spoiled. In the next quarter of a century, however, four developments combined to make bottled beer big business.

Pasteurization. The mystery of spoilage ended with Pasteur's *Studies on Beer,* published in 1876. The Father of Pasteurization discovered that fermentation (which, if unchecked, leads to spoilage) could be controlled through heat. Though his studies of fermentation went back as far as 1857, Pasteur only started concentrating on beer in the 1870's. Disheartened by the outcome of the Franco-Prussian War, his goal was to make French beer superior to that being brewed by the Germans.

Transportation. There was not much sense (or profit) in setting up large-scale bottling operations until there was a transportation system that would provide efficient and wide distribution. The great growth of the transcontinental railroad lines after the Civil War resolved this problem.

Sensible Laws. Up until 1890 the Internal Revenue Law made no provision for bottled beer. Therefore, the beer first had to be kegged in order to affix and cancel the necessary tax stamps, and then it had to be transported across a public road to a separate bottling facility. The cost of operating in this ridiculous manner was, of course, quite high and many brewers naturally limited their production to draft beer. In 1890 and 1891, however, common sense prevailed, and Congress passed laws enabling the brewer to pipe his beer directly from the brewhouse to the bottling plant without going through the barreling process. This change in governmental policy was a real boon for brewery bottling.

Acceptable Stopper. Through the last half of the nineteenth century over 1,500 different bottle-stopping methods were tried by U. S. brewers. None of them was very effective; all allowed dirt in and carbonation out. In 1892, however, William Painter invented the crown cap. This was actually a cover rather than a stopper, but it proved to be the effective bottle closing device that the brewing industry so desperately needed.

• • •

While all of the above factors played a major role in the progress made by bottled beer, the invention of the crown top is of special importance to today's bottle collector. Early bottles, whether glass or pottery, were handmade and generally quite crude, no two being 100 percent alike. The almost universal adoption of the crown top in the late 1890's and early 1900's, however, was the first step in conformity that led to machine-made bottles, automatic bottling and capping equipment and, eventually, complete bottle uniformity.

Prior to the advent of the crown top most beer bottles were of the "blob top" variety. While the advantages of the crown were recognized almost immediately, installation of the necessary capping equipment was expensive and it was therefore many years before the brewing industry had com-

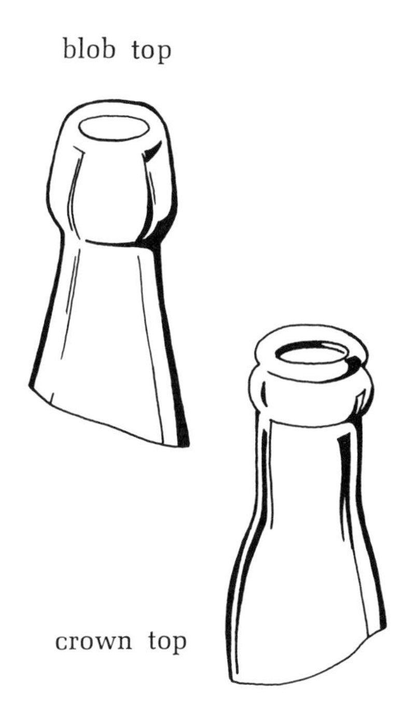

pletely made the switch. It is a definite mistake, therefore, to consider 1892 (or even 1900) as the hard and fast dividing line between the blob and the crown. On the contrary, some breweries used blobs well into the 1900's.

Turn-of-the-century bottling staff at the West End Brewing Co., Utica, New York.

No information appears to be available on pre-Prohibition embossing rather than labeling. Collectors, however, believe that many breweries never did use embossed bottles or else used them very sparingly. Early advertisements and trays picturing labeled bottles strongly support this belief.

One factor that limited the use of embossing was the expense; the brewer had to order a sizeable quantity from the bottle manufacturer in order to qualify for the special embossing. And then, despite the "This Bottle Must Be Returned" or "This Bottle Not to be Sold" statement found on virtually all embossed beers, they would disappear. Bottle bootlegging became quite widespread. Hotels, bars and anyone (especially young boys) who could get their hands on the bottles would sell them to a junkman or second-hand shop, which in turn would resell them back to the brewery. The brewery, of course, had no choice but to pay the price (usually quite steep) or else go out and order some new ones.

• • •

Beer bottle collectors travel, for the most part, in different circles from other breweriana collectors. While flea markets, antique shops and collector's publications are still valuable "sources of supply," the best source is bottle shows and clubs. Virtually every area of the country now has its own local bottle club, and most of these clubs hold one or more shows (which are actually bottle flea markets) a year. For information on bottle clubs in various areas, and possibly for a subscription, contact the *Old Bottle Magazine*, Box 243, Bend, Oregon 97701.

Beers, as beer bottles are known in the collecting world, have remained rather low in price relative to most other bottle-collecting categories. Good full-face embossed blobs at $2 or $3 each can still be found at many shows and shops east of the Rockies. On the West Coast prices for beers (as with virtually all other types of breweriana) are quite a bit higher.

"Full-face" embossed means there is raised lettering over a large percentage of the center part of the bottle (its "face"), as opposed to shoulder or bottom-of-the-bottle embossing. While labels can be extremely ornate, the embossed bottle has remained the favorite. This embossing, as we hope you'll notice in succeeding pages, can be extremely attractive. To insure readability, all of the lettering shown in illustrations to follow has been painted, with the exception of "Registered" or "This Bottle Not to be Sold."

Prior to the turn-of-the-century, bottling was often handled by someone other than the brewer. As mentioned, Federal regulations made brewery bottling difficult. Many brewers, themselves, firmly believed that brewers should brew and bottlers should bottle. As a result, early beers often carry, as illustrated, the name of a soda bottler or beer bottling house as well as the brewery's name.

A Comparison.

Bass' Ale is never bottled by its brewers, but is sold in bulk to independent bottlers, who affix to the bottles their own distinguishing labels.

Evans' Ale

Is Bottled at the Brewery

by its makers, who know precisely when and how to handle it, and who preciously guard the goodness of the brewing against danger from outside bottling. If it is not just right it is not bottled; that's why it's always brilliant and clear and absolutely **without sediment** and the

Best Ale in the World.

Some breweries, of course, did handle their own bottling, and the number increased rapidly after 1891. These do-it-yourself brewers often made promotional use of this fact, as can be seen in the 1903 *Life* (the original *Life*) ad placed for Evans' Ale. Even the bottles carry a statement drawing attention to the fact that they came from the brewery's own bottling works or bottling department.

Back when many "commercial" breweries were little more than enlarged home-brew operations, the embossed letters sometimes gave no hint that the bottle contained beer. These were all beers, but you'd never know it from the embossing.

Even more informal are these bottles—no town is given. The brewers were probably extremely local, and therefore didn't really have to include the town name; everybody knew it already.

Sets of bottles from the same brewery can add an interesting touch to any collection. Here are four sets that progress from name only to full-fledged "Brewing Co." status.

Essential to national success as a "shipping" brewery was a finely organized and tuned system of branches, agents and depots. This was one of the vital keys to the eventual standing of such present-day giants as Pabst, Schlitz and Anheuser-Busch. However, as shown here, many smaller breweries also had such systems, although regional rather than national. A typical example of the latter was probably the Anchor Brewing Co. of Dobbs Ferry, N. Y. Previously known as Peter Biegen's Brewery (1852-1881) and the Hudson River Brewing Co. (1881-1886), it operated as Anchor from 1886 until 1900. During this time it maintained a depot downriver in New York City at 11th Avenue and 33rd Street, and upriver in Poughkeepsie, where Henry Harris, a wholesale malt and liquor dealer, was the agent.

PICNIC BOTTLES

These are giant half-gallon beers that have come to be known as "picnics," undoubtedly because it was at such affairs that their large size came in handy. While *labeled* picnics have been found from breweries in all parts of the country, it appears that *embossed* picnics were predominately a Midwest special. Represented here are four states, with Iowa breweries accounting for 50% of the total bottles pictured.

The painted label Meister Brau was an early 1970's attempt to revive the picnic size. How successful it could have been will probably never be known as Meister Brau, a major Chicago brewery for many years, went out of business in 1972.

PRINTED LABEL BOTTLES

Post-Prohibition only, painted label (or applied label, as they are known in the brewing industry) bottles are quite collectible because only a relatively few breweries have used them.

Answering a question as to why and when a printed-label bottle has been used at all, Larry R. Levy, merchandising manager for the Reading Brewery, of Reading, Penn., writes, "We use it only on one package, the eight-ounce Pale Reserve line. We started using this in approximately 1937 and have used it on this one item ever since. The package is sold in bars by the bottle and through distributors by the case."

Both advantages and disadvantages in use of an applied label are explained by Levy. "On one hand, the initial cost of the bottle (which is returnable) is quite a bit higher than a blank one on which a paper label is used. In production, however, you save not only the cost of the label itself, but by-pass the labeling operation as well." The disadvantages, accordingly, are that "you are limited to a maximum of two colors and if you should want to change the design, other than throwing the old bottles away, you are stuck with a mixture for years to come. Tavern owners generally do not like applied labels as they prefer to put all empty bottles of one size from any brewery in an empty case and this is acceptable. With applied labels, this cannot be done."

With disadvantages seemingly outweighing advantages, the rarity of this type of bottle has become well established.

Remember the song "99 Bottles of Beer on the Wall"? Add 115 more and you have the number of beer bottles that are pictured, A to Z, over the next six pages.

BOTTLES, A to Z

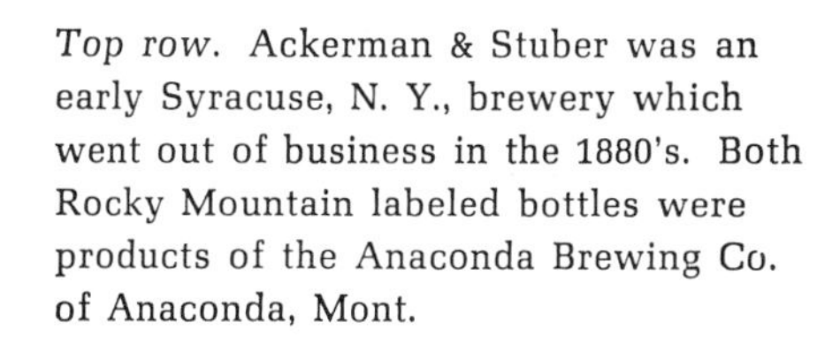

Top row. Ackerman & Stuber was an early Syracuse, N. Y., brewery which went out of business in the 1880's. Both Rocky Mountain labeled bottles were products of the Anaconda Brewing Co. of Anaconda, Mont.

Second row. Between 1905 and 1909 the George Bechtel Brewing Co. (third from right) of Stapleton, Staten Island, merged with the Bachmann Brewing Co. of Rosebank, Staten Island. Bechtel's plant was closed and the merged company operated in Rosebank as the Bachmann-Bechtel Brewing Co. (third from left).

Bottom row. The middle bottle, Collier Bros., is an English one. The embossing on this is about twice as pronounced as on American beers; whether this is true of all English beers of the period is not known.

Top row. Many breweries combined an ice business with brewing since (before the days of artificial refrigeration) they had to have ice on hand to properly lager the beer. This explains the Crystal Spring Brewing & Ice Co. embossing.

Second row. Dukehart's Maryland Brewery was well known for its malt tonic, the probable contents of this unusually shaped bottle.

Third row. Finlay (second from right), a famous brand in Ohio for many years, was one of the real pioneer breweries of Toledo. It was founded by William J. Finlay in 1853.

Bottom row. For most of the pre-Prohibition period, the Florida Brewing Co., founded in 1897, was the only brewery in the entire Sunshine State. Ironically, because of artificial refrigeration and population increases, Florida is now, in the 1970's, a rather significant brewing state. Fuhrmann & Schmidt, founded in 1860 in Shamokin, Penn., is still in operation, though it has been a subsidiary of Ortlieb since 1967.

Top row. Both the Globe and Golden Age half-gallon "kegs" appear much older than they actually are. Globe, a San Francisco brewer, was in business for only five years, 1934-1939. And the Golden Age Brewery of Spokane, Washington, didn't last much longer, opening its doors in 1934 and closing down in 1948.

Second row. Grace Brothers was a familiar name in California for close to a hundred years. Founded in 1873, as a steam beer brewery, the company operated for 93 years, until 1966.

Bottom row. The Home Brewing Co. was a small firm that began business in 1891, and remained in operation until 1969; Richbrau was its best known brew. On the Honolulu B & M (Brewing & Malting) bottle notice the "T. H." which stands for the Territory of Hawaii. This brewery was founded in 1898 and operated until 1918.

Top row. If you look closely you can see part of the stopper in the Jamestown Brewing Co. bottle.

Second row. Lake City Brewery (far left) was the early name of the Fred Koch Brewery. It is still very much in operation in the Western New York city of Dunkirk.

Third row. The McAvoy Brewery (second from left) made many friends when the Chicago fire occurred in October of 1871. The fire destroyed the city waterworks, leaving virtually the entire city (including the McAvoy plant) without water. Within three days, however, McAvoy's General Manager, H. Bemis, laid a pipe directly to Lake Michigan, erected pumping equipment, and supplied water to the brewery and the citizenry as well. Thousands of people reportedly lined up their wagons before the brewery to take on a supply of water.

Top row. Robert Portner (far right) more or less backed into the brewing business. He started a grocery store in Alexandria soon after the beginning of the Civil War. Alexandria, directly across the Potomac from Washington, was the encampment site for many Union soldiers, and was teeming with Pennsylvania-German troops, men who naturally wanted beer to ease the burdens of their Army duties. As the war progressed, however, it became harder and harder to keep the soldiers supplied, as most of the brew was brought down from New York and Pennsylvania. So, Portner, in the fall of 1862, started to brew his own. He went on to become an extremely successful Virginia brewer.

Third row. As well as providing Syracuse a good beer, brewer Thomas Ryan also served the city as mayor for three terms.

Bottom row. One of the East Coast's best known breweries, F. & M. Schaefer, is represented here at the 51st and 4th Avenue address in New York City that it called home for 67 years, from 1849 to 1916. In that year, due to increasing land values on 4th (now Park) Avenue and the need for expansion, Schaefer moved to its present location in Brooklyn.

Top row. The small bottle that reads "Brewers" is a very rare sample bottle from the Springfield Brewing Co. of Springfield, Mass. Two bottles in this row include the brewery's telephone number, a rather unusual practice. The two are K. G. Schmidt of Chicago and the Star Brewing Co. of Washington, Penn.

Second row. The labeled Volga bottle, still filled with beer, is from the United States Brewing Corp. of Red Bluff, California (1942-1948).

Third row. Old Sleigh half and half was a product of the Victor Brewing Co. (between 1905 and 1909 to 1941) of Jeannette, Penn. The label has a copyright date of 1936.

Bottom row. Roger Williams Ale could only have been brewed in Rhode Island, and it was, by the Roger Williams Brewing Corp. of Providence, in operation from 1934 to 1940.

Trays

If breweriana in general is a colorful galaxy among twentieth-century collectibles, then trays are the brightest stars in that galaxy. No breweriana collection is really complete without at least a few trays to add color and grace. And once a few are found, others are sure to follow.

The brewing of beer itself goes back over 6,000 years, but an important number of packaging and advertising innovations are purely American. The beer can and crown cap come readily to mind. So should the beer tray, for it was in Coshocton, Ohio, that the tray was conceived in the very late nineteenth century. As small a town as Coshocton was, it had two rival specialty advertising companies. The first of these, Tuscarora Advertising, was organized by J. F. Meek in 1887. A year later the second company, Standard Advertising, was founded by H. D. Beach. Each firm tried to outdo the other with respect to the types of advertising items produced—burlap school bags, umbrellas, chair backs, fans, grocery aprons, calendars, rulers, cloth caps, thermometers, booklets, etc., etc. You name it and if it somehow could contain an advertising message either Tuscarora or Standard (or both) produced it.

Eventually this kind of very healthy competition led to the development of lithography on metal, first on signs and then serving trays. Though the exact date is unknown, the first advertising trays were probably produced in the early to mid-1890's. They've been manufactured ever since, of course, and by numerous other companies. A list of the more prolific of these manufacturers, with their appropriate dates of operation, is given in the back of the book. Since many trays include the manufacturer's name (usually in small print near the center of the bottom inside rim), these dates of operation are very helpful in determining age. Also helpful is the copyright date that appears on some trays, generally in the same bottom inside rim area. This is given when the manufacturer (or, more rarely, the brewery) thought enough of the tray's design to protect it. While a copyright date doesn't insure that the tray was manufactured that year, one can be sure it was produced within a year or two of the date (and certainly not before it).

While the attempt to date trays is interesting, it is a definite mistake to consider only really old (i.e., pre-Prohibition) items as collectible. While pre-1919 trays are, admittedly, much more uniformly attractive, at the rate breweries have gone (and are going) out of business virtually all trays are very collectible. Just as important as the age is the beauty and the condition of the tray.

There are two basic tray sizes—12″ diameter and 13″ diameter. Many of the really outstanding pre-Prohibition trays, however, are oval in shape, and a fair percentage are square or rectangular. Another entire category of trays, not used for serving at all, is that of miniatures. These are more accurately known as change or tip trays, as the waiter or barmaid used it to leave the customer's change and, hopefully, to receive a tip. Generally 4″-5″ in diameter, these little beauties are avidly sought by many collectors, and, in fact, often sell for more than a serving tray with the same design.

Many of the more ornate scenes or portraits used were manufacturer's stock designs; i.e., the scene was standard and the brewery ordering the tray simply specified copy. This, of course, made their purchase more economical, and explains why the "same tray" was used by several different breweries. Stock or brewery's own, there are several themes that have predominated in all tray designs through the years. Over the next few pages a number of examples of each of these themes are shown.

COMIC TRAYS

Trays with cartoon-type scenes or happy-go-lucky characters have also been used by breweries over the years. This seemed to have been especially true in Providence, R. I., where Hanley's famous "Connoisseur" (see page 130) was matched by Providence Brewing Co.'s comical gentleman.

A magnifying glass is needed to count the number of people in the Gunther Gambrinus tray (bottom center). Rumor has it that there are 19, including King Gambrinus himself.

The "Who Wants the Handsome Waiter" sign includes a classic design made available by the Hampden Brewing Co. (now Piel Bros.) of Willimansett, Mass., in 1934. The same design appeared on their beer tray; thus the sign is included here with other trays.

1934 is also the copyright date on the upper right Gunther tray, and probably the Ruppert Knickerbocker tray, upper left, is of the same vintage.

Otto was an appropriate name for the harried waiter on the Erlanger tray, for it was the Otto Erlanger Brewing Co. His brewery, unfortunately, did not last very long; it opened in 1939 and closed in 1951.

Since this change tray is hardly visible above, we've included a close-up here. Phil. Schneider was a small brewery, located at Plum and Convent Streets in Trinidad, Colo. One of the earliest of this state's breweries, Schneider can trace its beginnings back prior to 1878. It went out of business in 1941. Brands were LaFiesta, Century, Silver State, Zephyr and, of course, Schneider's.

GUNTHER'S
BEER
WHO WANTS THE HANDSOME WAITER
Gretz
BEER
OTTO SAYS:
THIRST
COME...
THIRST
SERVED!
ERLANGER
deluxe BEER
GAMBRINUS
KING OF LAGER

METAL TRAYS

The fact that breweries often spared no expense when producing advertising and promotional material is illustrated here. These are all solid metal trays. The first two are of a silver-plated metal of some variety, while the last four are brass. With many of today's trays being made of plastic, we've "progressed" from solid brass to solid plastic in less than 100 years.

While interesting to study in full light, these trays do not display or photograph well. Therefore, below each tray the wording that is included on it is given. "A.B.C." did not stand for Anheuser-Busch Co. A.B.C. was the American Brewing Co., a large St. Louis firm located at 2815 S. Broadway. American was best known for its bottled beer, and had a large foreign export business, which explains why the legend reads, St. Louis, U.S.A. instead of the usual St. Louis, Mo.

Jac. Kiewel Brewing Co.
The Home of White Rose
Little Falls, Minn.

Phoenix Brewery
Celebrated Pearl Bottled
and Lager Beer
St. Louis, Mo.

King of all Bottled Beers
A.B.C.
St. Louis, U.S.A.

Diogenes Brewing Company
Wyckoff Ave. & Decatur St.
Brooklyn Boro, N. Y.
Lager Beer

L. Bergdoll
Brewing Co.'s.
Lager Beer
Philada. Pa.

Terre Haute Brewing Co.
Brewers & Bottlers
of High Grade Beers
Capacity 450,000 Barrels
Indiana's Leading Brewery

ENAMEL TRAYS

Remember the old enamel or porcelain license plates that some states used back in the very early days of motoring? These trays were made in exactly the same fashion—a thick coating of enamel over a base of metal. They are heavy and quite rare, apparently being used only by Eastern United States and Canadian breweries.

All the American trays shown here were manufactured by the Baltimore Enamel & Novelty Co., of Baltimore. They also had a branch at 190 West Broadway, New York City.

Carefully read the Geo. Brehm & Son tray in the bottom row and you'll catch the week's "shopper's special"—$1.00 for a case of 24 bottles. What later became Geo. Brehm & Son was originally the George Neisendorfer Brewery. Neisendorfer founded it in 1858, and operated the brewery for seven years before his death in 1865. About that time George Brehm came into the picture and, after a brief courtship, married Mrs. Neisendorfer in May of 1866. When the name of the company was officially changed is unknown, but it was as George Brehm (and later, starting in 1899, as Geo. Brehm & Son) that this company developed into one of Baltimore's better known breweries.

VIENNA ART

Vienna Art plates, strictly speaking, are not trays, but they are part of the tray "family," and are usually collected and displayed in the same manner. They were manufactured by the H. D. Beach Co., successor to Standard Advertising, of Coshocton, Ohio, and all that I've seen bear a patent date of Feb. 21, 1905 on the reverse side.

Vienna Art plates generally picture a girl on the front, and the advertising message, if any, is found on the reverse. Several of those shown here read on the back, "Anheuser-Busch's Malt-Nutrine, St. Louis." The lower left plate, a restaurant tray, states, "Heilich's, 1319 Arch St., Opp. Elk's Home, Philadelphia. The Louis Bergdoll Brewing Co.'s Beer Exclusively."

A pleasant exception to the standard motif is found on the tray from The Frank Jones Brewing Co., a very sizeable and respectable New England ale and porter brewery. Frank Jones, himself, in addition to being a most successful brewer, was also very active in New Hampshire politics and was a United States Congressman from the Granite State for two terms, 1875-1879.

ORNATE LETTERING

The beauty of most trays is such that they do not have to include a pretty girl, a brewery scene or an animal vignette. The lettering alone can make them stand out in any display.

George Ehret deserves special mention. At one time this country's largest brewer, he emigrated to the United States from Germany in 1857 at the age of 22. For quite a few years Ehret worked in the Hupfel brewery on 161st St. in New York. In 1866 he struck out on his own and founded the famous Hell Gate Brewery (named for the section of New York in which it was located) at 235 East 92nd Street. Right from the start the brewery did extremely well. In 1871 production was 33,512 barrels; two years later, in 1873, it had more than doubled to 74,497 barrels; and by 1877 Ehret's was the largest brewery in the country, with production of 157,825 barrels. And in the years following, the brewery did even better: 220,096 barrels in 1880; 292,899 in 1885 and 412,851 in 1890. With sales going so well, however, Ehret apparently did not wish to expand much beyond the New York metropolitan area, and did not involve himself in a far-flung agency and depot system, as did several of the larger Midwestern brewers. Therefore, in spite of Ehret's own leaps and bounds, Pabst far outdistanced him by 1890, with sales of over 1,000,000 barrels. Schlitz and Anheuser-Busch had also pulled ahead by this time.

With the coming of Prohibition the brewery turned to the manufacture of cereal beverages. After Repeal they reopened, but at 193 Melrose Street in Brooklyn. The company operated at this address until 1948 and then moved to New Jersey, taking over the William Peter Brewing Co. plant at 3315 Hudson Street in Union City. This operation, however, lasted but a few years, and by 1951 this once world famous brewery had joined the ranks of the out-of-business.

For more information on Ehret's post-Prohibition Brooklyn brewery see page 173.

TRAYS, A to Z

On the next four pages are illustrations of 61 trays, arranged alphabetically by brewery. Comments have been held to a minimum so that the trays themselves can be given as much space as possible.

Two Dawson's trays (second row from the bottom, extreme left and right) are included. One is an up-dated version of the other, with different hair styles, furnishings, clothing, etc. I believe the left-hand tray was issued soon after Repeal, and that on the right c. 1940.

Breweries have always vied to come up with the catchiest slogans. On this page, for example, is found "The Beer With a Cheer" (Kuhn's); "You Get More Out of Hampden" (Hampden); "Pride of the Valley" (Lebanon Valley); "Brewed Today Grandfather's Way" (Hensler); "Drink the Best" (Jung's); and two "Still the Best" (Hedrick and Horlacher). Horlacher won this contest, as they are still in operation, while Hedrick, Albany's last remaining independent brewery, ceased operations in 1966.

The gentleman pictured on the Hensler tray (top row, extreme right) is, incidentally, none other than Joseph Hensler, one of New Jersey's pioneer brewers.

Notice the tray on the extreme left, second row down from the top. This is from a Panamanian brewery. Even in Central America they knew the selling power of the word "Milwaukee."

Two trays deserve mention. One is the tray in the middle of the second row from the bottom. It is unusual in that it was put out to commemorate a special event—the coronation of King George VI on May 12, 1937. The brewery issuing this tray was, of course, English—Frederic Robinson's Unicorn Brewery in Stockport.

The second unusual tray is that on the left of the second row from the top, "Rochester Made Means Quality." This was paid for and distributed by Rochester's four post-Prohibition breweries, American, Standard, Genesee and Rochester. A rare example of solidarity among competing companies.

Old Advertisements

As a brewer began to think about expanding his market—be it into a different county, state or even a distant part of the same city—he began to think about advertising. Early ads, usually in local newspapers, were little more than announcements. "We are proud to announce that Stenroos Lager is now available at Anderson's Beer Hall." That sort of thing.

Gradually, brewers became more sophisticated in their advertising. If volume warranted it, many eventually hired agencies to write the bulk of their sales copy. By 1903, for example, Pabst was utilizing the services of the J. Walter Thompson agency.

While few pre-Prohibition breweries used large circulation magazines, several did. These four ads are all from turn-of-the-century mass magazines.

Evans Ale was once so well known that it could be stated in 1914:

Turn it
Upside Down
Drink it all.
There are no dregs.

The success of 100 years of brewing embodied in

Evans' India Pale Ale.

The drink for all who love good things.

Rich as Cream.
Without Sediment.
Free from False Ferments.
Allowed two years to ripen in the Wood before bottling.

At all Summer Resorts.

C. H. EVANS & SONS
Brewery & Bottling Works,
HUDSON, N. Y.

England boasts that the sun never sets upon her flag; The Evans' of Hudson can boast that the sun never sets upon their Ale. This product of Hudson's oldest industry is enjoyed in many languages and by many nations.

The ale and stout which are manufactured and marketed by the house of Evans have added to Hudson's reputation at home and abroad. Many a distant port knows of Hudson only as the home of the ale that is brewed in this city. A bottle of it on the table in a strange land is as sympathetic a bond of greeting as the mystic grip of any lodge or the secret pass word of any fraternity.

—excerpt from a September, 1914 pamphlet, "Atlantic Deeper Waterways Association at Hudson, N. Y."

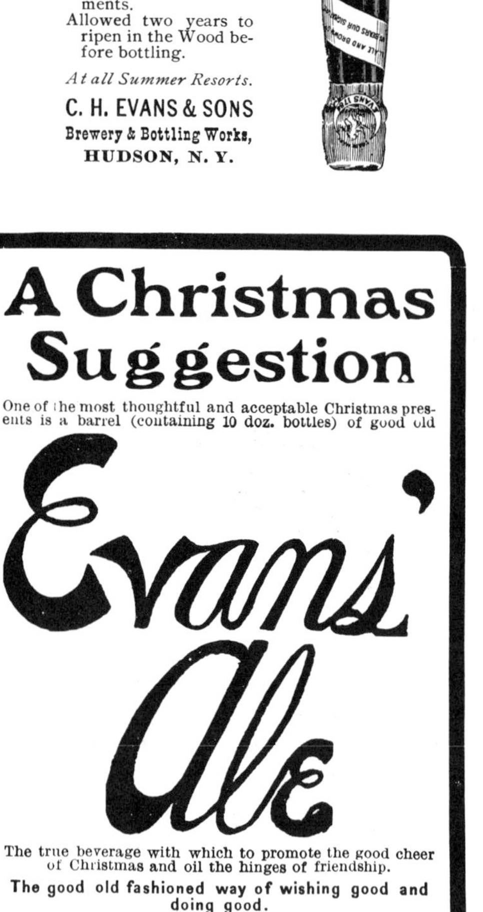

Evans never reopened after Prohibition, although the Evans name was used by the Peter Barmann Brewing Co. of Kingston, N. Y., until its demise in 1941.

Do you Drink Beer?

If so, the chief points to consider are *flavor*, *body*, *color*, *digestive properties*.

A beer which excels in all these qualities is essentially *healthful*, *invigorating* and *refreshing*.

Epicures Say that BEADLESTON & WOERZ'S **Imperial Beer** surpasses all other beers on this account. Competent judges pronounce it to be the **King of Beers.** Wherever it is introduced it immediately becomes popular.

Experience proves that **Imperial Beer** is the best table beer in the market.

Tests Show that it is the best beer for the tropics, thus demonstrating its "keeping qualities."

Highest Awards and Diplomas received in open competition *testify* to the high standard of excellence maintained by

Imperial Beer.

The leading hotels and clubs keep "Imperial," because it is always in demand.

BEADLESTON & WOERZ, N. Y. City.
All first-class Grocers will supply you.

A pretty girl—can you think of a better way to catch a man's attention? Probably not, and apparently neither have American breweries and their advertising agencies, for this is a motif that has been consistently used since beer advertising was first begun. For probably the same reason—there is no better way to catch the eye—trays that picture a pretty girl are generally considered the most collectible of all, and command high prices.

Most of these are stock trays, and the same girl can therefore be found advertising several different brands of beer. Some of these young ladies were even given names by the tray manufacturers. The girl on the upper left tray (Star–Union Brewing Co., Peru, Ill., 1856-1966) is "Janice," and the model next to her on the Harry F. Bowler tray is "Mildred."

The Highlander Beer tray (middle row, right) is from the Garden City Brewing Co., of Missoula, Montana. Garden City was a pre-prohibition brewery only, operating from 1896 to 1919. The Highlander brand name, however, was kept alive by the post-prohibition Missoula Brewing Co., 1933-1964.

A picture of the product being promoted is a very basic advertising device—both to increase brand-name awareness and to instill a desire to try the product. And, with beer, this has been especially effective; who can resist the desire to have a brew once you've seen a sparkling cold, foamy glass of beer.

On the top left and right are two extremely colorful trays from Mèxico.

Hauenstein was a small-sized brewery that produced good Minnesota beer until 1970. John Hauenstein, the company's founder, more-or-less backed into the business. In 1862 he was working for a New Ulm distillery that was almost completely destroyed in a Sioux Indian attack. Finding himself unemployed (and perhaps longing for a taste of old Bavaria, his birthplace), Hauenstein joined in partnership with Andreas Betz in 1864 to found the brewery that bore his name for over 100 years.

Here's a well-decorated bottle of beer. The Indianapolis Brewing Co. took the grand prize at the 1904 St. Louis World's Fair for their Gold Medal Düsseldorfer. The brewery was so overjoyed at receiving the great honor that it closed the plant for a day and put on a spectacular parade for the benefit of the people of Indianapolis.

Animals and animal scenes are quite often pictured on trays. This was sometimes due to a brewery using the name of an animal as a brand name. Red Fox (a product of the Largay Brewing Co., Waterbury, Conn.), and Pony Deluxe are perfect examples of this practice. More often, however, animals were used solely to highlight and beautify the tray, with no concern about a logical tie to the brand name or brewery.

Moose, horses and eagles seem to have been the symbols most often used, but beer trays also have pictures of lions, tigers, dogs, roosters, pelicans and chipmunks—in fact, just about every species of animal life.

Along with pretty girls, the other tray "category" favored by many collectors is the brewery scene. These objects, by definition, were not stock trays. Each was produced solely for the brewery illustrated.

The J. Chr. G. Hupfel Brewery, located at 223-229 East 38th St. in New York City, as it was portrayed prior to Prohibition on a change tray. Next to the tray is the same building as it looks today, the home of Century Moving and Storage Co.

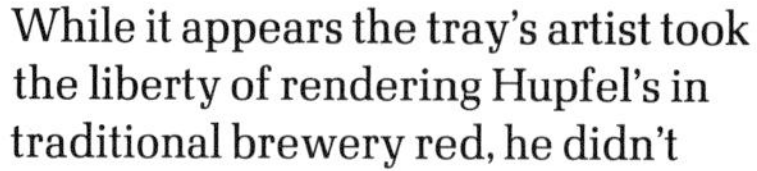
While it appears the tray's artist took the liberty of rendering Hupfel's in traditional brewery red, he didn't use his imagination. The building WAS red and somewhere along the way was painted its present color.

The Tannhauser Double Brew (or, if you prefer, Resuahnnat Werb Elbuod) illustration is a proof sheet for a tray that was produced for the Consumers Brewing Co., of New Orleans. Consumers was an extremely short-lived firm, opening shortly before Prohibition, between 1905 and 1909, and closing for good in 1919.

The Ebling tray (top illustration, bottom row, left) is courtesy of America Hurrah Antiques, New York, N.Y.

Spout tops, the most collectible of all cans.

FIRST ROW: Atlantic Ale, Atlantic Co., Atlanta, Ga./Kessler, Kessler Brewing Co., Helena, Mont./Silver Bar, Southern Brewing Co., Tampa, Fla./National, National Brewing Co., Baltimore, Md./ Rocky Mountain, Anaconda Brewing Co., Anaconda, Mont./Highlander, Missoula Brewing Co., Missoula, Mont./ Ehret's Extra, George Ehret Brewery, Union City, N.J.

SECOND ROW: Red Fox Ale, Largay Brewing Co., Waterbury, Conn./Fort Pitt, Fort Pitt Brewing Co., Pittsburgh, Pa./Cincinnati Burger Brau, Burger Brewing Co., Cincinnati, O./Moosehead, Moosehead Breweries, Lancaster, N.B., Canada/Spearman, Spearman Brewing Co., Pensacola, Fla./Little Dutch, Wacker Brewing Co., Lancaster, Pa./National Bohemian, National Brewing Co., Baltimore, Md.

THIRD ROW: SB, Southern Brewing Co., Tampa, Fla./Tropical, Tampa Florida Brewery, Tampa, Fla./Ortlieb's, Henry F. Ortlieb Brewing Co., Philadelphia, Pa./Ziegler's, Louis Ziegler Brewing Co., Beaver Dam, Wis./Old England, Old England Brewing Co., Derby, Conn./Stoney's, Jones Brewing Co., Smithton, Pa./Iron City, Pittsburgh Brewing Co., Pittsburgh, Pa.

FOURTH ROW: Burkhardt's, Burkhardt Brewing Co., Akron, O./Koehler's, Erie Brewing Co., Erie, Pa./Augustiner, August Wagner Breweries, Columbus, O./Wooden Shoe, Wooden Shoe Brewing Co., Minster, O./Aero Club, East Idaho Brewing Co., Pocatello, Ida./ Regal, American Brewing Co., New Orleans, La./National Ale, National Brewing Co., Baltimore, Md.

While old flat-top cans do not have the collecting glamour of their spout-top cousins, they can be extremely colorful.

First row: Pfeiffer's, Pfeiffer Brewing Co., Detroit, Mich./Krueger, G. Krueger Brewing Co., Newark, N.J./Schmidt's City Club, Jacob Schmidt Brewing Co., St. Paul, Minn./both Hull's, Hull Brewing Co., New Haven, Conn./Famous Black Dallas, Leisy Brewing Co., Cleveland, O./both Gold & Silver Mug, Lebanon Valley Brewing Co., Lebanon, Pa.

Second row: Stoeckle, Diamond State Brewery, Wilmington, Del./Best Beer, Best Brewing Co., Chicago, Ill./Imperial, Southern Brewing Co., Los Angeles, Cal./Prager, Atlas Brewing Co., Chicago, Ill./Dixie, Mountain Brewing Co., Roanoke, Va./Drewry's Ltd., U.S.A., South Bend, Ind./Sheridan, Sheridan Brewing Co., Sheridan, Wyo./Atlantic Ale, Atlantic Brewing Co., Charlotte, N.C.

Third row: Miller, Miller Brewing Co., Milwaukee, Wis./Balboa, Southern Brewing Co., Los Angeles, Cal./Gunther, Gunther Brewing Co., Baltimore, Md./Straight 8, Spearman Brewing Co., Pensacola, Fla./Hapsburg, Best Brewing Corp., Chicago, Ill./Diamond State, Diamond State Brewery, Wilmington, Del./Great Falls Select, Great Falls Breweries, Great Falls, Mont./Schoenling, Schoenling Brewing Co., Cincinnati, O,

Fourth row: Old Dutch, International Breweries, Findlay, O./Narragansett, Narragansett Brewing Co., Cranston, R.I./Muehlebach, Geo. Muehlebach Brewing Co., Kansas City, Mo./Burkhardt's, Burkhardt Brewing Co., Tacoma, Wash./Pearl, San Antonio Brewing Co., San Antonio, Tex./Waldorf Red Band, Forest City Brewery, Cleveland, O./ Fort Pitt, Fort Pitt Brewing Co., Pittsburgh, Pa.

Holidays were occasions for celebration—Washington's Birthday, New Year's, Valentine's Day—and breweries took advantage of high spirits.

While the Reuter & Co. Highland Spring Brewery poster is not a Valentine's card, it is fanciful enough to have been used for amatory purposes. The 1882 Val. Blatz card reflects the long-standing German tradition of celebrating New Year's Day. The Barthlomay Washington's Birthday card is, however, impossible to explain. The birthday of the Founding Father has never been, to common knowledge, a holiday for which cards have been sent. Nevertheless, it is a most attractive piece of Americana.

Trade cards were very popular in the United States in the late 1800's and early 1900's. Millions of them were distributed as a form of advertising by a variety of companies, both large and small. Their use by breweries, however, was quite limited.

While all the cards shown on this page were used for advertising, the only true trade cards are the two issued by the Phoenix Bottling Co. to promote H. Clausen & Son. Phoenix did Clausen's bottling.

The Lloyd Brewing Co. "card" was actually a sample label put out by George Schlegel, a New York City label manufacturer, to demonstrate his skill. As such it served as his "calling card." Lloyd was a fictitious brewery.

Local and state business directories are great sources for very early brewery ads.

CHAS. MOAT & CO.,

BREWERS

—OF—

Pale, Amber & Stock Ale,

WASHINGTON STREET (near Freight Depot),

CHAS. MOAT,
JAMES WADMAN,
WM. J. MOAT, BREWER.

AMSTERDAM, N. Y.

EVANS'

HUDSON ALE,

BY

PHIPPS & EVANS,

HUDSON, N. Y.

JAMES L. PHIPPS, C. H. EVANS.

502 ALBANY DIRECTORY.

AMSDELL BROS.,

BREWERS OF

ALBANY XX ALE

CELEBRATED INDIA PALE

—AND—

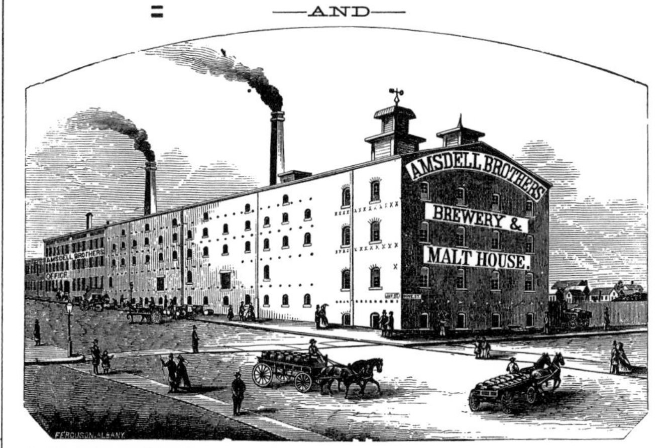

AMBER ALE & PORTER

AMSDELL BROS.' WORLD RENOWNED

DIAMOND and BURTON ALE.

Brewery and Malt House, JAY, DOVE and LANCASTER Sts., ALBANY, N. Y.

DEPOTS,

NEW YORK, 444 and 446 GREENWICH STREET....AMSDELL BROTHERS

Westfield, Mass	M. Donovan	Amsterdam, N. Y	W. P. Kline
Pittsfield, Mass	J. P. Rouse	Catskill, N. Y	W. J. Martin
Burlington, Vt	W. L. Stone	New Hamburgh, N. Y	John Eagan
New Brunswick, N. J	D. McCurdy	Troy, N. Y	John Shields
Williamsburgh, N. Y	J. H. Thieman	West Troy, N. Y	John Shields
Rondout, N. Y	J. H. Cullen	Cohoes, N. Y	John Shields
Fultonville, N. Y	G. G. Firth	Waterford, N. Y	John Shields
New York	J. J. Ward	Poughkeepsie, N. Y	M. J. Downey
New York	James Campbell	Mineville, N. Y	Denis Hayes
Newark, N. J	F. McGenis	Windsor Locks, Ct	P. J. Keating
Gloversville, N. Y	S. S. Gross	Oak Hill, N. Y	C. W. Osborn
Johnstown, N. Y	R. Windsor	Whitehall, N. Y	Jno. O'Neil
Schenectady, N. Y	Jonathan Levi	New Haven, Ct	T. J. O'Connell
Coxsackie, N. Y	E. D. Fancher		

64 *GEER'S BUSINESS DIRECTORY.* 505

HUBERT FISCHER,

Lager Beer Brewery,

ICE PLANT

—AND—

REFRIGERATING ROOM,

Extra Lager FOR BOTTLING a Specialty.

The De LaVergne System.

315 Park, cor. Lawrence St., Hartford, Conn.

HARTFORD BREWING CO.,

BREWERS OF

FINE CANADA MALT,

Fine Ales and Porter,

Flat, Stock, and East India Pale Ales a Specialty.

Wholesale Dealers in Malt.

Brewery, 232 Sheldon Street, cor. Front Street, Hartford, Conn.

PETER CHUTE, Proprietor.

Incorporated April 21, 1879. Capital, $37,500. *Address all communications to the Company.*

PETER CHUTE, *President.* OSCAR KOENIG, *Vice President.*
JULIUS HUEBLER, *Secretary.* WM. W. FRICK, *Treas. and Manager.*

Directors.—Peter Chute, Julius Huebler, J. Melrose, O. Koenig, Wm. McKone, Wm. W. Frick, Mrs. Charles Herold.

THE HEROLD

CAPITOL BREWING CO.

Brewery, Nos. 52 to 62 BELLEVUE ST., Hartford, Conn.

Extra Fine Lager Beer

SPECIALLY BREWED FOR BOTTLING.

STOCK ALE and PORTER.

Trade directories are an additional source of beer advertising. These two ads appeared in the April, 1909, issue of the *North American Wine & Spirit Journal,* published in Boston, Mass.

Narragansett, of course, continues in operation today as the Eastern arm of Falstaff.

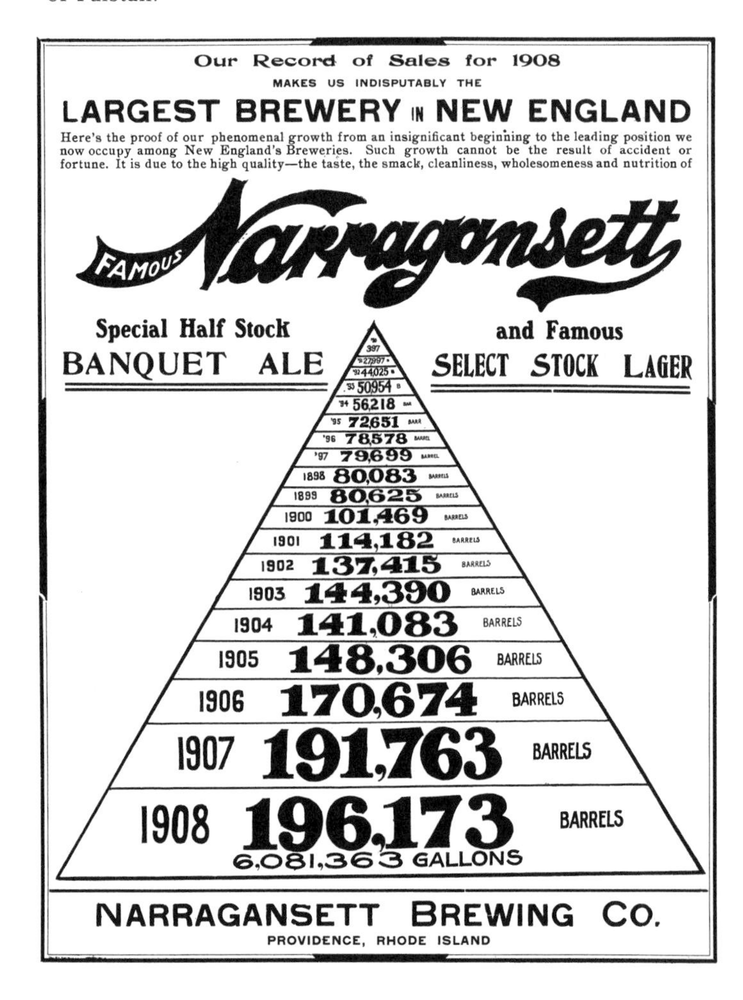

Notice the man admiring his glass of Granite State. A "beer belly" such as this is never found in contemporary beer ads. True Jones, by the way, was the brother of brewer Frank Jones.

Some fine old advertisements that appeared in an 1885/1886 directory with the unbelievable title of *The New York City Record of the Wholesale and Retail Wine and Liquor Dealers, Ale and Beer Brewers, Mineral Water Dealers, Segar and Tobacco Manufacturers of the City and County of New York.*

98 NEW YORK CITY RECORD.

GEO. WINTER BREWING CO.

Put up expressly for Export, Family and Club House use, by

A. WOLFF,

222 East 55th Street.

NEW YORK CITY RECORD. 131

FLANAGAN, NAY & CO.,

COLUMBIAN

Brewery and Malt House

No. 450 West 26th St.,

Between 9th and 10th Avenues,

JAMES FLANAGAN,
JOSEPH O. NAY,
WM. L. FLANAGAN.

NEW YORK.

132 NEW YORK CITY RECORD.

S. KRESS, President. H. M. HAAR, Vice-President and Treas.

THE JOHN KRESS BREWING CO.

211-221 AND 218-224 EAST 54TH STREET,

NEW YORK.

CHAS. GUNTHER, Secretary. H. GUNTHER, Superintendent.

PATRICK KIERNAN. JOHN KEENAN. AUGUSTUS T. DOCHARTY.

DAVID JONES COMPANY,

BREWERS OF THE CELEBRATED

Old Amber and Pale Stock Ales,

SUPERIOR XX AND XXX CANADA MALT NEW ALES.

BREWERY AND OFFICE,

N. W. Cor. First Avenue and 44th Street.

Calendars

Calendars were yet another item distributed free by brewers to taverns, hotels and restaurants. And for good reason. If a patron wanted to know the date he would check the calendar and, naturally, notice the ad for the brewery's beer. On most brewery calendars the dates themselves occupy but a very small part of the poster's available space. All that was probably visible in a dimly lighted taproom was the ad, which, of course, is a fact that makes these old calendars very collectible today. Most are quite ornate and extremely colorful.

Many calendars feature scenes of the brewery itself, as in the case of Ebling, a Bronx brewery founded by brothers Philip and William Ebling in 1868. This plant was used until the closing of the business in 1951. The building is still standing in an area of the Bronx badly hit by urban blight, and the splendor that once radiated from this brewery radiates no more.

This calendar was printed as a duotone. Many were in full color, however, as can be seen in the color pages.

Two fine Philadelphia calendars, and an outstanding piece of calendar art. Notice the heavenly look on the face of the Rieger & Gretz girl, the implication being that Rieger & Gretz brew has a divine taste. The boy pictured in the 1912 Vollmer calendar appears a little young to have carried a bottle of beer in his lunch basket; perhaps it was for his father.

J. & M. Haffen was a Bronx brewery founded in 1856 by Mathias Haffen. Before his success in brewing, Haffen had pursued quite a variety of occupations–government employee, railroad construction worker and Long Island farmer. The company's featured brand was Neffah Brew. If you spell Neffah backwards, you'll see how he came up with such an unusual name.

Tap Knobs

The next time you visit a neighborhood pub, look over the draft beer area and you're bound to see several tap knobs. Federal law requires that the beer's brand name be displayed so that a customer knows what he's getting.

A "hot-rodder" of the 1940's and 50's often used one of these knobs on his gearshift. Old cars and trucks, believe it or not, are a good source of vintage knobs. Beer distributors and breweries themselves, of course, are the main source, but usually only for the newer tap varieties.

Tap knobs make an interesting collectible because they are colorful, come in an infinite variety of shapes, and are small enough to be readily displayed. In the past few years, however, some breweries have been using much larger and more ornate taps.

While most tap knobs will stand by themselves, a surer way is to mount them, as has been done with many of the knobs pictured here.

These almost appear to be miniature trays, but they are the enamel inserts from early tap knobs. These, probably because of manufacturing over-runs, turn up more often than you'd suspect at flea markets and antique shops. Virtually all of the inserts shown on this page, for example, were found in an upstate New York antique shop framed and hung on the wall.

"As if in tribute to beer's antiquity—and popularity—man through the ages has devised special vessels from which to drink the beverage. In fact, no other beverage has inspired such a wide variety of drinking containers. They range from wood, leather, and pottery to gold, silver, pewter, porcelain, and glass—from beakers, bowls and tankards to steins, tumblers, fancy goblets, and paper cups. Beer drinking vessels have been utterly crude and supremely ornate. They are beautiful and grotesque, practical and supercilious."

—John R. McBride, in "Trifles and Treasures"

Glasses and Mugs

While any vessel that properly holds beer can be deemed practical,· this section deals with only the very practical—those that were used in advertising beer as well as holding it.

American breweries started issuing advertising glasses in the early 1880's. Embossed glasses, á la the Anheuser-Busch goblet and the Virginia Brewing Co. glass mug, came first and were in vogue until around the turn-of-the-century. At that point etched glasses started to become popular, followed by heavy pre-Prohibition use of frosted glasses.

The large pitcher in the upper left is a real beauty. It's solid copper and reads Graham's XXX Ale. On the reverse side is a vertical series of small circles, covered with glass, to allow the user to see how much beer is left. This is most likely from the old Graham Ale Brewery in Paterson, N. J.

La Tropical was a product of the Tampa Florida Brewery, located in that city. Before Prohibition this brewery was known as the Florida Brewing Co.; in 1934 it reopened as Tampa Florida Brewery, and remained in operation under that name until 1960.

While these two types appear virtually identical, etching is a process that cuts into the glass while frosting is a fancy term for spraying onto (but not into) the surface. Whether etched or frosted, most pre-Prohibition glasses had white lettering. Color has been predominantly a since-Repeal phenomenon.

Even more recently, of course, we have the traditional painted glass. Here, too, the paint is sprayed on but with no attempt to achieve the hazy or frosted look. Examples are Pioneer, Holihan's, Dominion, and La Tropical, plus numerous others shown here. In addition to being spray painted and featuring color, post-Prohibition glasses tend to be much thicker than their turn-of-the-century counterparts.

Advertising mugs were introduced roughly around the same time as glasses, in the last quarter of the nineteenth century. The most notable characteristic of American mugs is the small size when compared with their European counterparts. While this may be partially explained by the liter/pint difference, the major reason is that Americans prefer their beer much colder than Europeans, and have thus rejected large mugs or tankards for fear of lukewarm brew toward the bottom of the container.

The Fesenmeier item pictured in the upper left hand had two uses—first, as a Christmas candle and, when finished, as a large goblet. Fesenmeier was West Virginia's last operating brewery, going out of business in 1971. It was known as the Little Switzerland Brewing Co. for the last four years of its existence.

To the right of the candle-goblet is an Arrow 77 glass. Arrow 77 was originally a product of Baltimore's Globe Brewery. The Arrow part of the name was added in 1920. Globe was searching for a clever name for its Prohibition near beer, and a contest was sponsored, offering $1,000 to the person who submitted the best idea. "Arrow Special—It Hits the Spot" was the winner.

The shiny mug in the upper right is also post-Prohibition, and was a fairly short-lived product of The Revere Copper Co., of Rome, N. Y. Known as a "Tapster," it was developed to "tap" a can of beer. To operate, the lid is opened, the cold can is inserted in the mug, and the lid closed. Believe it or not, you can then pour. The inside has a built-in opener, and is positioned so that the now-opened can pours through the spout.

Pabst Harlem was not a branch or agency, but a restaurant. And what a restaurant it was! When it opened in September of 1900 it was the largest restaurant in America. 1400 could be seated at one time. And they were probably all drinking Pabst—which is the reason so many breweries were also in the restaurant, saloon and hotel business prior to Prohibition.

My favorite mug is the miniature Bartholomay in the right forefront. But, unfortunately, I know nothing about it or the July, 1911, FEEL-TER-HUM that it commemorates (or even what a FEEL-TER-HUM is or was).

A nice grouping of all pre-Prohibition (with the exception of Pioneer) glasses. As you can see, there was no uniform shape when it came to glasses and goblets. The Piel's Real German Lager (upper left) looks, as a matter of fact, as if it should hold an ice cream sundae instead of beer.

Health Articles

Beer, in addition to being a beverage of moderation, has always been considered a wholesome and healthful product, supplying both carbohydrates and protein to the body. It is also highly recommended as a tonic—easing tension while providing nutrition. In fact, it would be difficult to think of another beverage that can so effectively both relax and refresh.

Early beer advertising and packaging placed a fair degree of emphasis on these health aspects. This, of course, is in sharp contrast to today's ads which tend to promote beer more for its *social* virtues than for its *health* benefits.

This early crockery bottle has to represent the height of ornateness. *Old Jug Lager* is described as "fully matured, a veritable luxury. The fashionable beverage of the day. Brilliant in color, absolutely pure, stimulating, rejuvenating."

Even the name of the company was squeezed in—The Moerlein-Gerst Brewing Co., Nashville, Tenn. U.S.A. This brewery, while in business for many years prior to Prohibition, was known as Moerlein-Gerst for only three years, 1890 to 1893.

This product of the German-American Brewing Co. of Buffalo is described as "Pure Food Beer." It is also "Sterilized, not pasteurized!"

These five containers run the gamut from health to sickness—from Park Brew and Henry Zeltner which, since they're clean and pure, will keep you that way, to the not very positive approach of Mueller Bros., "No Drugs or Poison," which means that all they guarantee is that their beer won't kill you. Hohen-Adel's Health Beer would, however, presumably cure you of any ill effects, as would Gowdy's Medicated Beer.

Some other examples of pre-Prohibition "health" ads.

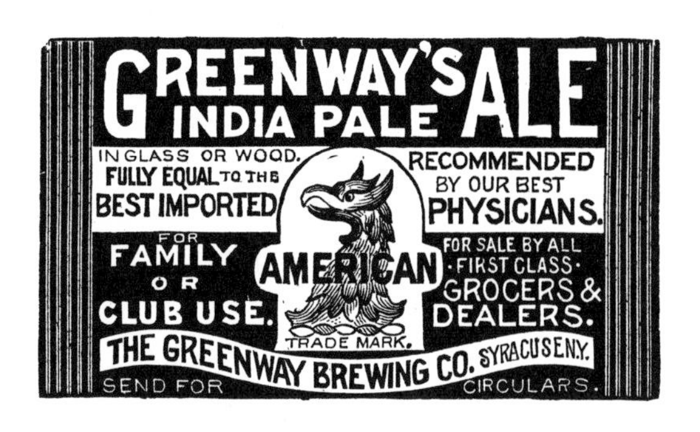

MUHLENBERG BREWING COMPANY

BREWERS OF BEER AND PORTER

A Product that Has
Stood the Test for

QUALITY
AND PURITY

Recommended by
Leading Physicians

READING, PA.

A Bottle of Beer

providing the beer is good, is the most healthful and refreshing of drinks. When the beer is bad, the drinker knows how extremely unpleasant are the after effects.

GOOD BEERS excel in *flavor*, *color*, *body*, and *digestive qualities*. They are free from excess of gas and all deleterious admixtures. They promote digestion and benefit the health. Beers conforming to this high standard are few in number. There is only one which surpasses it and that is

Imperial Beer.

It becomes popular wherever it is introduced. Competent judges say it is

THE BEER TO DRINK.

The leading hotels and clubs keep the "Imperial" because it is always in demand.

Experience proves that **Imperial Beer** is the best table beer in the market.

Tests show that it is the best beer for the tropics by reason of its wonderful keeping qualities.

Highest Awards and Diplomas received in open competition *testify* to the high standard of excellence maintained by **Imperial Beer.**

BEADLESTON & WOERZ,

NEW YORK CITY.

All first-class Grocers will supply you.

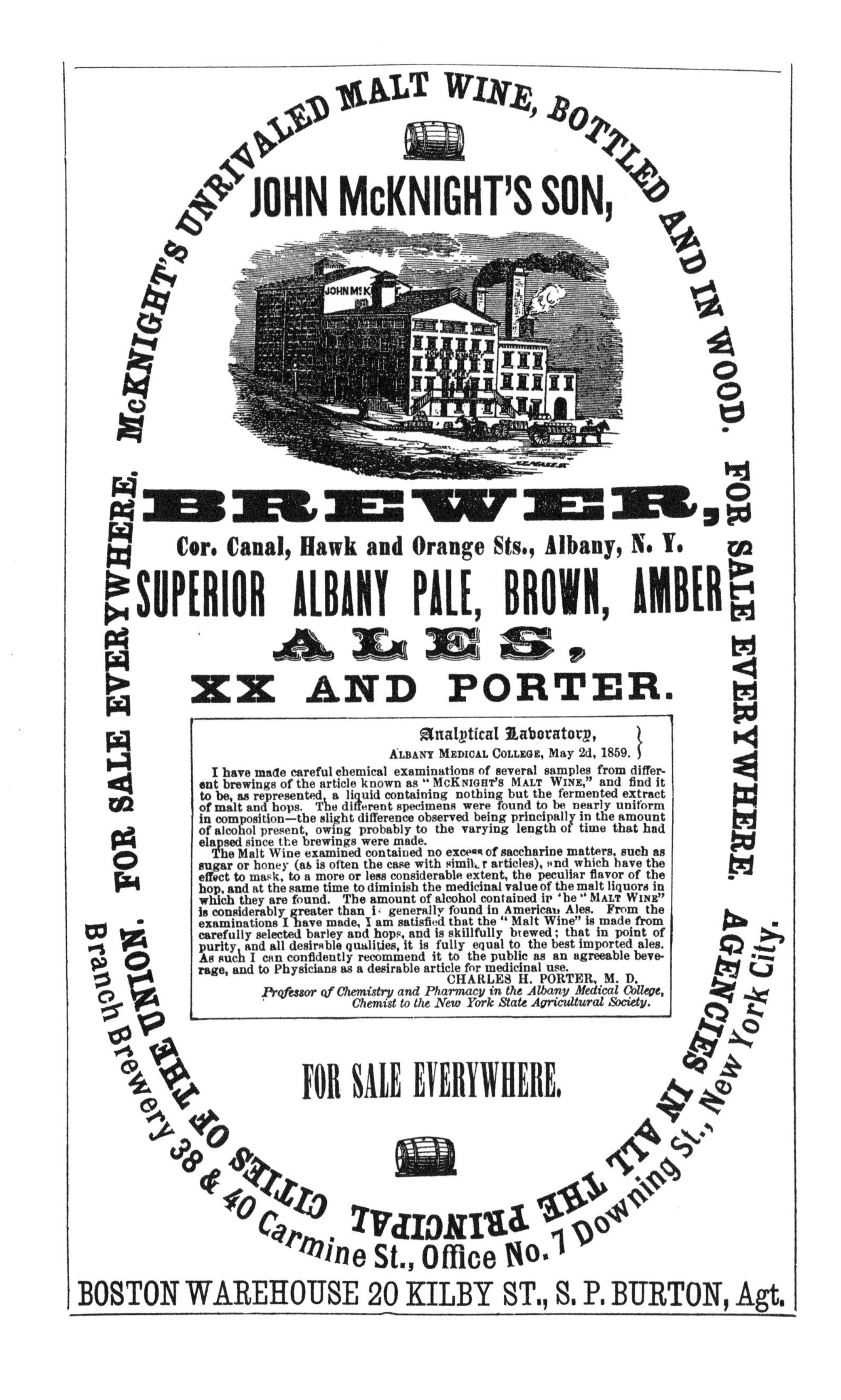
McKNIGHT'S UNRIVALED MALT WINE, BOTTLED AND IN WOOD.
JOHN McKNIGHT'S SON,
JOHN McK
BREWER,
Cor. Canal, Hawk and Orange Sts., Albany, N. Y.
SUPERIOR ALBANY PALE, BROWN, AMBER
ALES,
XX AND PORTER.
Analytical Laboratory,
Albany Medical College, May 2d, 1859.
I have made careful chemical examinations of several samples from different brewings of the article known as "McKnight's Malt Wine," and find it to be, as represented, a liquid containing nothing but the fermented extract of malt and hops. The different specimens were found to be nearly uniform in composition—the slight difference observed being principally in the amount of alcohol present, owing probably to the varying length of time that had elapsed since the brewings were made.
The Malt Wine examined contained no excess of saccharine matters, such as sugar or honey (as is often the case with similar articles), and which have the effect to mask, to a more or less considerable extent, the peculiar flavor of the hop, and at the same time to diminish the medicinal value of the malt liquors in which they are found. The amount of alcohol contained in the "Malt Wine" is considerably greater than is generally found in American Ales. From the examinations I have made, I am satisfied that the "Malt Wine" is made from carefully selected barley and hops, and is skillfully brewed; that in point of purity, and all desirable qualities, it is fully equal to the best imported ales. As such I can confidently recommend it to the public as an agreeable beverage, and to Physicians as a desirable article for medicinal use.
CHARLES H. PORTER, M. D.
Professor of Chemistry and Pharmacy in the Albany Medical College,
Chemist to the New York State Agricultural Society.
FOR SALE EVERYWHERE.
FOR SALE EVERYWHERE. FOR SALE EVERYWHERE.
FOR SALE EVERYWHERE. AGENCIES IN ALL THE PRINCIPAL CITIES OF THE UNION.
Branch Brewery 38 & 40 Carmine St., Office No. 7 Downing St., New York City.
BOSTON WAREHOUSE 20 KILBY ST., S. P. BURTON, Agt.

A Pure Beer for the Whole Family

A natural food product, every drop laden with body building, healthful food substances.

PETER DOELGER

FIRST PRIZE BOTTLED BEER

"Expressly for the Home"

Is veritable bottled energy, absolutely free from germ life and impurities of any kind.

A little higher in price than ordinary beer—a great deal higher in quality

Supplied by all first-class dealers. Served in leading Hotels and Cafes

BOTTLED EXCLUSIVELY AT

PETER DOELGER FIRST PRIZE BREWERY

Bottling Dept., 407-433 East 55th Street - - New York

51

For better or worse, this card very graphically illustrates the health benefits of Wiedenmayer's Beer! Wiedenmayer was a Newark, N. J., brewer.

I DRINK GEO.W. WIEDENMAYER'S BEER

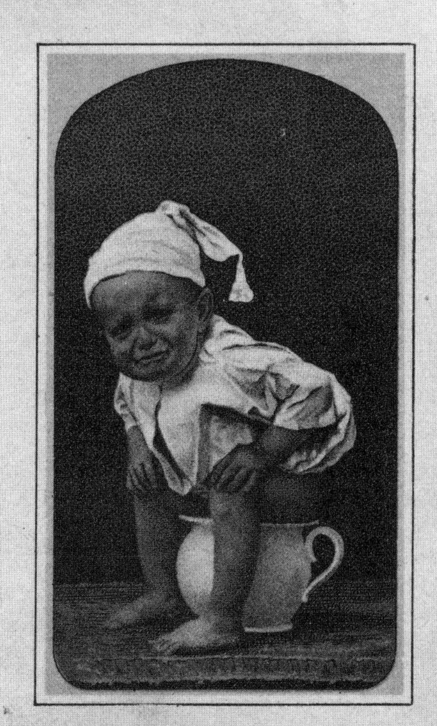

I DID'NT.

OVER

Cans

The beer can is the newest category of breweriana, and yet it is probably the most collected. There are possibly more collectors of cans than of trays, bottles and glasses combined! And there are good reasons for this. First, cans are inexpensive and easily obtainable. You can purchase a different brand each time you buy a six-pack. Second, in spite of the high brewery mortality rate, there are still hundreds of American brands. Most of them are distributed in a purely local or regional area, making travel fun for the collector. Third, and perhaps most important, cans are extremely colorful and possess a surprising degree of beauty when properly displayed.

The beer can was born in 1935. In January of that year the G. Krueger Brewing Co. of Newark, N. J., became the first brewery in the world to market beer in cans. The site selected for the marketing test was Richmond, Va., a city large enough to provide a meaningful sample, and yet far out on Krueger's marketing periphery, just in case the whole idea flopped.

And there were undoubtedly many industry people who thought Krueger's test would flop. The beer can had been in the "test tube" stage for a number of years, and no one seemed capable of solving all the problems involved in the seemingly easy transition from food-in-cans to beer-in-cans. Some of these problems were caused by the much greater internal pressure produced by beer, a highly carbonated beverage. Far and away the biggest problem, however, was the fact that beer is an extremely sensitive product, and is

especially "allergic" to metal. It was necessary, therefore, to develop a super lining to serve as insulation between the beer and the actual metal container.

The American Can Company (Canco), working since 1931, was the first company to claim that they had solved these problems. Their can was flat-topped, of three-piece heavy steel construction, and featured a special enamel lining. Their can was trademarked "Keglined" in September of 1934, and the company set out to find a brewery that would give it a trial run under actual "battlefield" conditions. Krueger, after several months of deliberation, became American's first customer, agreeing to the test in Richmond.

The rest is history. Within two months beer in cans was a smashing success in Richmond, and Krueger had increased its share-of-market considerably.

Such success did not, of course, go unnoticed by Krueger's fellow brewers. Pabst, for example, unveiled its "TapaCan" in July of 1935, and Schlitz introduced its beer in cans two months later. The Schlitz can, however, was quite different from the one being used by Krueger and Pabst. Instead of being flat, it had a crown-sealed spout top, very reminiscent of a bottle. This can, known as "Cap-Sealed," was the Continental Can Company's entry in the field. At first, the design appeared to have several advantages. It did not need a special opener, as did the flat-top. It was quite similar to a bottle in shape, and therefore seemed more familiar to beer drinkers. And, most importantly, the spout can could be filled with traditional bottling machines, thus making it unnecessary to invest in expensive canning equipment. For this last reason, many smaller breweries leaped on the "Cap-Sealed" bandwagon.

The real advantage of the beer can, however, was its compactness. Lighter and smaller than bottles (and non-returnable, too), it brought beer transportation costs way down. Of the two can styles, however, one was much more compact than the other. And that, of course, was the flat-top can. A beer truck could hold many more cases of flat-tops than spouts. And retail outlets preferred them for much the same reason—they stacked better and took less space. In addition, the flat-top turned out to be much more adaptable to high-speed filling equipment than the spout, with its smaller opening. As a result of these very real disadvantages, the spout can, as produced by Continental and later Crown Cork and Seal, had a relatively short life span—from 1935 until the mid-1950's.

While the can had real advantages from the viewpoint of the brewer and the retailer, it had to be accepted by the beer-drinking public if it was to become more than a short-lived fad. As a result, in the period from 1935 to 1940 the can was promoted through the most expensive advertising campaign ever conducted for any kind of container. In addition, all early cans carried statements or even lengthy passages extolling the virtues of canned beer.

Twenty-seven years after the beer can was first introduced in Virginia, the Old Dominion State again served as the testing ground for a beer-packaging

Examples of early can "sell" copy.

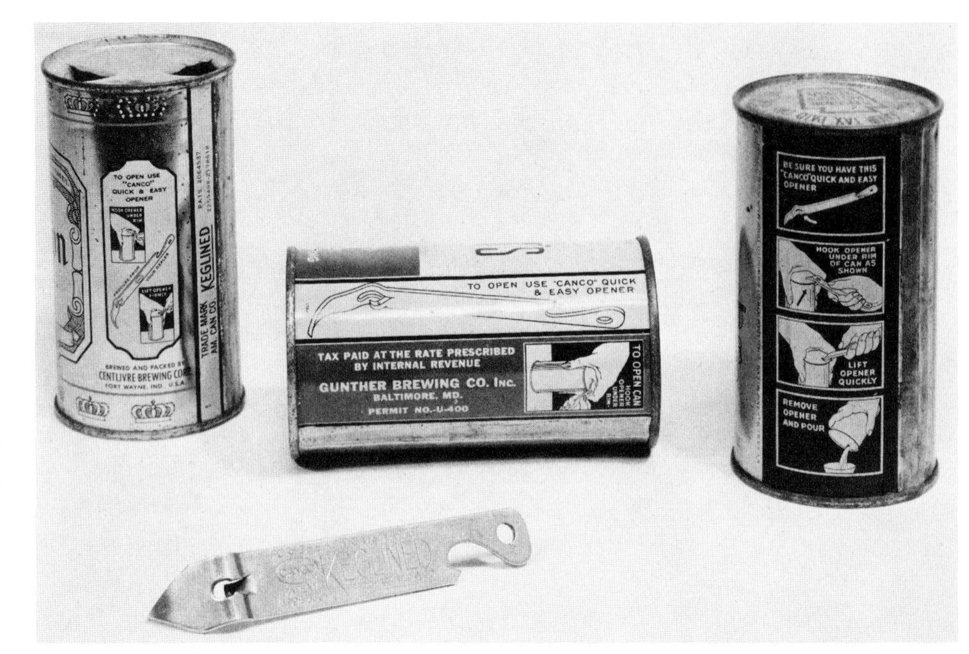

Another problem that had to be overcome was the fact that, as strange as it may sound today, opening a beer can was a strange new experience that required a strange new device–the opener. The American Can Company developed the "Quick and Easy" punch-type opener, and licensed the right to manufacture and distribute it to the Vaughan Novelty Manufacturing Co. These openers, both with and without specific brewery advertising, were distributed free by the tens of thousands in order to insure that every red-blooded, beer-drinking American had easy access to one or more openers. Instructions on how to use an opener were given on almost all pre-World War II cans. Notice that the set of instructions provided on all three cans shows only one hole punched. It apparently took a while to realize that beer flows faster and better with two punches. Also pictured is an original style Canco Keglined opener. It is 5″ long and, somewhat surprisingly, is made to open bottles as well as cans. In the 1950's, due to rising costs, most give-away openers were shortened to approximately 3″.

innovation, the tab top opener. Developed by Alcoa Aluminum for use with its aluminum tops (in use since the late 1950's), this revolutionary openerless-opener was first used by the Pittsburgh Brewing Co. for its Iron City beer in 1962. A year later Schlitz began using the tab top and promoted it heavily through television, newspaper and billboard advertising. With brewing the competitive industry it is, other breweries could not afford to be one step behind, and within the year more than 65 brands of canned beer had tab or lift tops. By 1965, 70% of all cans produced had some type of easy-open device.

Use of the beer can has had great significance in this country. From the brewers' viewpoint the can has had very dramatic effects. One, it greatly stimulated total beer sales, and, two, it greatly helped develop a national market.

Although Prohibition fell far short of its goal of a dry society, it did curb greatly the alcoholic intake of most of "middle America," that large amorphous mass that consumes so much of the "brewer's art." Worse yet, most young people did not have the chance to acquire a taste for beer–and beer is definitely an acquired taste beverage. As a result of these factors, plus the effect of the Great Depression, per capita beer consumption in the first full year following Repeal was but 50% of what it was in 1919, the last year before Prohibition. The beer industry needed something new and dramatically different to boost sales, and the beer can, probably more than any other single ingredient, provided that boost. Ironically, though, the comeback came in a round-about way. Although the can had helped stimulate sales prior to World War II, by 1941 only 186 of the 507 United States breweries in operation were using cans, and cans accounted for but 14% of total packaged beer sales. With the war, restrictions were placed on tinplate and the beer can disappeared from the domestic market. Because of non-breakage and compactability, however, cans continued to be utilized for military shipments. During the war, in fact, over one billion beer cans were shipped to American fighting forces around the globe. And when "Johnny Came Marching Home" again he, naturally, wanted to continue drinking his beer from a can. Drinking from the can has been perfectly proper ever since.

A strong secondary effect of use of the can was to further strengthen the already strong position of the national giants. Even though they could afford the very latest and most efficient machinery and gained prestige through heavy advertising campaigns, the nationals had one big disadvantage in their constant battle with local and regional brewers—heavier transportation costs. With the arrival of the can, however, shipping charges were lowered substantially and they could compete even more effectively with the locals. In fact, I would hazard a guess that, if the beer can had never been invented, there would probably still be several hundred American brewers in operation, instead of the present 70.

Lastly, and perhaps most important, the beer can has had significant ecological or environmental effects. Prior to its introduction, virtually all malt liquors and soft drinks were packaged in returnable bottles. The can's success, in other words, spearheaded much of today's "throwaway" attitude among both packagers and consumers. Without getting involved in what is often a very emotional issue, it goes without saying that many ecology-minded organizations in this country look upon the tin throwaway with nothing but disdain. In fact, in what may prove to be pioneer legislation, the State of Oregon has attempted to strongly curtail the use of all non-returnable cans and bottles. In 1972 it became the first state to (a) require a deposit on *all* beer and carbonated-beverage containers and, (b) outlaw the use of the tab top. Since the law went into effect the beer can has virtually disappeared in Oregon. With 36 other state legislatures considering similar statutes, the future of the can, and can collecting as we know it today, could be in serious jeopardy.

While it is possible for the collector to find still-produced brands and can designs, it can be difficult and challenging to find older containers in collectible condition. This situation is a result of the can's seeming worthlessness. Trays, glasses, signs and many other forms of breweriana are often hoarded away in old bars, hotels and even by the average person who holds on to them because they seem to be something of value. And eventually these hoarded items surface at auctions, estate sales or flea markets. And bottles, actually the oldest type of breweriana, are literally surfaced most every sunny weekend when bottle diggers take to the old dumps. But cans are something else. For every million produced, 999,999 have been thrown away. As a result, finding a mint pre-World War II spout can is considerably more difficult than finding a mint pre-World War I tray, glass or sign.

Because of this rarity, early cans are slowly working their way into the antique category and, as such, are beginning to show up at flea markets and antique shops. Eventually they should take their rightful place in the antique world beside tobacco tins and early spice, tea and coffee canisters.

NOVELTY CANS

In the relatively few years since 1935 numerous novelty or special interest cans have been produced. Some of these are highlighted on this and the next page.

Two decorative Japanese sake cans. What makes these worthy of interest, of course, is the attached miniature opener. And it really works, although you get a mighty small punch.

While beer can cigarette lighters are still available today, they obviously are not of the spout-top variety. Both of these were Bower's Peli-cans, guaranteed to "give you more lights from one filling than a Siamese lightning bug on his honeymoon!"

When was the last time you saw a beer can with a price printed on it? Probably never, because it completely cuts out pricing flexibility. These two cans are from the Maier Brewing Co., Los Angeles, Calif.

These are truly special-event cans, and are unusual in that they are dated. MBAA stands for Masters Brewers Association of America. This can was issued in 1963 when MBAA had its 76th Anniversary convention at the Lord Baltimore Hotel in Baltimore. The "1,000,000-man hours without a lost time accident" can was made by Carling in honor of its Natick, Mass., safety achievement. The two middle cans once must have been the property of a returning Tiger alumnus.

In spite of strict quality control, incorrectly or incompletely printed cans occasionally avoid detection. The Mark V can to the left, missing its second color, was found amidst a six-pack of fully printed cans (exemplified by the right-hand can). Mark V is a product of August Wagner Breweries, Columbus, Ohio.

As with trays, pretty girls have adorned many a beer can. Here's a set pointing up the prettiest girl of them all—Miss Olde Frothingslosh.
The story is that a mid-1960's Pittsburgh disc jockey used to herald the Thanksgiving/Christmas holiday season by announcing that it was time for some Olde Frothingslosh. This private joke was supposedly repeated for a number of years until, lo and behold, one year Olde Frothingslosh, "the pale stale ale from the foam on the bottom," suddenly appeared on the Steel City scene during the holiday season. At first this special brew was packaged only in bottles, with comical labels. However, in 1969 the Pittsburgh Brewing Co. proclaimed a Miss Olde Frothingslosh, and issued the special can pictured here to commemorate this great event. Space prohibits including all the fine information included on the reverse of the can, but we'll cover a few of the more important points. For example, you're probably curious how Fatima Yechburgh, the girl that won it all, was selected. She "was chosen on the basis of beauty, talent, poise . . . and quantity." Where do beauty queens come from? Fatima's "from a small town outside Pittsburgh. It's considerably smaller since she left." Lastly, "Miss Frothingslosh's formula for success: Think Big."

Culture also comes with beer. Introduced in July of 1952 by the Globe Brewing Co. of Baltimore, Md., Hals beer cans and labels featured "The Laughing Cavalier," a celebrated painting by Franz Hals.

Make the beer can interesting and sell more beer! That undoubtedly was the philosophy behind these three containers. Esslinger parti quiz contained a whole series issued in the early 1960's, each can including little known facts to stump your drinking companion. A rival Philadelphia brewery, Wm. Gretz, also had a series, "Tooner Schooner."

These were used during the 1950's and featured a well known song on each can—great for a then-popular sing-a-long. Similar to Gretz's "Tooner Schooner" was the Hitt's Sangerfest can issued by the Walter Brewing Co. of Pueblo, Colo., in 1966 and 1967. True to its name (a Sangerfest is a song festival) this can included the words to five songs.

Gretz had another series in the 1950's which revolved around G. B. (Gretz Brewing) fleet cars. In addition to the MGA shown here, the series included the Fiat 600 Multipla and Austin Healey 100.

This may be the most interesting of all novelty cans. Probably some kind of soda can, it reads, "no fl. oz." and "infernal revenue nix paid," (a take-off on "Internal Revenue tax paid," a statement included on all cans sold in the United States from 1935 until 1950).

GALLON CANS

Your own real draft beer at home. Sounds good—and in the mid-1960's the concept also appealed to quite a few American breweries. Developed by National Can, the idea unfortunately never really took hold with consumers. While of great interest to collectors, these gallon cans were heavy to handle, had to be kept refrigerated at all times, and did not keep well once the can had been tapped. It was probably just too much bother and too much beer for most American households.

Fifteen of the ill-fated gallon cans. Because they were produced for only five or six years, these cans are already quite rare.

SPOUT-TOP CANS, A to Z

While flat-top cans actually came first, it is the spout can that has caught the fancy of most can collectors. Consequently, 219 spouts are included here and over the next four pages. Included are numerous quart spouts and several World War II olive drab military cans.

First row: both Tam O'Shanter Ale, American Brewing Co., Rochester, N. Y./ Atlantic, Atlantic Co., Atlanta, Ga./ American, American Brewery, Baltimore, Md./Ballantine's Export, P. Ballantine & Sons, Newark, N. J./Sunshine, Barbey's Inc., Reading, Penn./both Barclay's Export, Barclay Perkins & Co., London, England/Bavarian Old Style, Bavarian Brewing Co., Covington, Ky.

Second row: Berghoff, Berghoff Brewing Corp., Fort Wayne, Ind./both Old Fashion and Billings, Billings Brewing Co., Billings, Mont./all three Blackhawk, Blackhawk Brewing Co., Davenport, Iowa/both Blatz, Blatz Brewing Co., Milwaukee, Wis.

Third row: Breidt's, Peter Breidt Brewing Co., Elizabeth, N. J./Vat Reserve, Brewery Management Corp., New York, N. Y./ both Buckeye, Buckeye Brewing Co., Toledo, Ohio/Buffalo, Buffalo Brewery, Sacramento, Calif./all three Burger, Burger Brewing Co., Cincinnati, Ohio/ Butte Special, Butte Brewing Co., Butte, Mont.

Bottom row: Canadian Ace, Canadian Ace Brewing Co., Chicago, Ill./both 12-oz. Carling's, The Carling Breweries, Ltd., Waterloo, Ont., Canada/both quart Carling's, Brewing Corp. of America, Cleveland, Ohio/Castletown, Castletown Brewery, Castletown, Isle of Man, England/Regent, Century Brewing Corp., Norfolk, Va./Mexicali, Cerveceria De Mexicali, Mexical, B.CFA, Mexico.

First row: all three Union and Malta Caracas, Cerveceria Union, Caracas, Venezuela/Chester, Chester Brewery, Chester, Penn./Tally-Ho, City Brewing Corp., New York, N. Y./Deluxe, Corona Brewing Corp., San Juan, P. R./Croft, Croft Brewing Co., Boston, Mass.

Second row: Croft, Croft Brewing Co., Boston, Mass./all three Old Export, Cumberland Brewing Co., Cumberland, Md./Black Horse Ale, Dawes Black Horse Brewery, Montreal, Quebec/Dawson's, Dawson's Brewery, New Bedford, Mass./ Diamond State, Diamond State Brewery, Wilmington, Del./Diehl Five Star, Diehl Brewing Co., Defiance, Ohio.

Third row: Diehl, Christ Diehl Brewing Co., Defiance, Ohio/Dixie 45, Dixie Brewing Co., New Orleans, La./Dow, Dow Brewery, Montreal, Quebec/first two Duquesne, Duquesne Brewing Co., Pittsburgh, Plant #2, Stowe Twp., Penn./ third Duquesne, Duquesne Brewing Co. of Pittsburgh, Penn./both Ebling's, Ebling Brewing Co., New York, N. Y.

Fourth row: both Ebling's (paper over metal cans, very rare) and Michel, Ebling Brewing Co., New York, N. Y./Valley Brew, El Dorado Brewing Co., Stockton, Calif./Clyde Cream Ale, Enterprise Brewing Co., Fall River, Mass./both Koehler's, Erie Brewing Co., Erie, Penn./ Erlanger, Otto Erlanger Brewing Co., Philadelphia, Penn.

Fifth row: Erlanger, Otto Erlanger Brewing Co., Philadelphia, Penn./Fehr's, Frank Fehr Brewing Co., Louisville, Ky./Wolf's, Fernwood Brewing Co., Lansdowne, Penn./West Virginia Special, Fesenmeier Brewing Co., Huntington, W. Va./all five Fitzgeralds, Fitzgerald Bros. Brewing Co., Troy, N. Y.

First row: first four Fort Pitt, Fort Pitt Brewing Co., Pittsburgh, Penn./right-hand Fort Pitt and both Old Shay, Fort Pitt Brewing Co., Jeannette, Penn./Ben Brew, The Franklin Brewing Co., Columbus, Ohio.

Second row: Silver Fox, Fox Deluxe Brewing Co. of Indiana, Marion, Ind./Fox Deluxe, Fox Deluxe Brewing Co., Grand Rapids, Mich./all three F. & S., Fuhrmann & Schmidt Brewing Co., Shamokin, Penn./Glasgo, Glasgow Brewing Co., Norfolk, Va./Graupner's, Robert H. Graupner, Harrisburg, Penn./Horton, The Greater New York Brewery, New York, N. Y.

Third row: all three Gretz, William Gretz Brewing Co., Philadelphia, Penn./ Hampden Ale, Hampden Brewing Co., Willimansett, Mass./Hauenstein, John Hauenstein Co., New Ulm, Minn./ Heineken, Heineken De Venezuela, Caracas, Venezuela/Richbrau, Home Brewing Co., Richmond, Va.

Fourth row: Richbrau, Home Brewing Co., Richmond, Va./both Hudepohl, Hudepohl Brewing Co., Cincinnati, Ohio/ Hull Brewery, Hull Brewery Co., Ltd., Hull, England/Hull's Beer and Hull's Ale, Hull Brewing Co., New Haven, Conn./ both Iroquois, Iroquois Beverage Corp., Buffalo, N. Y.

Fifth row: Jeffrey's, John Jeffrey & Co., Edinburgh, Scotland/Keeley Half & Half, Keeley Brewing Co., Chicago, Ill./White Seal and Grain Lager, Kiewel Brewing Co., Little Falls, Minn./Kingsbury, Kingsbury Breweries Co., Sheboygan, Wis./Krueger, Krueger Brewery Co., Wilmington, Delaware/Krueger Quart, G. Krueger Brewing Co., Newark, N. J.

First row: all three Kuebler, Kuebler Brewing Co., Easton, Penn./Rolling Rock, Latrobe Brewing Co., Latrobe, Penn./ Lebanon Valley, Lebanon Valley Brewing Co., Lebanon, Penn./Leisy's, The Leisy Brewing Co., Cleveland, Ohio/both Lowenbrau, The Lion Brewery, Munich, Germany.

Second row: Lucky Lager, Lucky Lager Brewing Co., San Francisco, Calif./ Pilser's Original and Old Dutch, Metropolis Brewery, New York, N. Y./ Moosehead, Moosehead Breweries, Lancaster, N. B., Canada/both Bavarian, Mount Carbon Brewery, Pottsville, Penn./National Bohemian, The National Brewing Co., Baltimore, Md.

Third row: Dawes Kingsbeer, National Breweries, Lt., Montreal, Quebec/John Bull and Old Bohemian, New Philadelphia Brewery, New Philadelphia, Ohio/both Tru Blue and Old Fashioned, Northampton Brewery Corp., Northampton, Penn./ both Oertel's '92, Oertel Brewing Co., Louisville, Ky.

Fourth row: Stag's Head Ale, A. Keith & Son, Halifax, N. S., Canada/Oland's Export Ale, Oland & Son, Halifax, N. S., Canada/Old Reading, Old Reading Brewery, Reading, Penn./both Ortlieb's, Henry F. Ortlieb Brewing Co., Philadelphia, Penn./Regal, Peoples Brewing Co., Duluth, Minn./P.O.S., Philadelphia Brewing Co., Philadelphia, Penn./ Pilser's, Pilser Brewing Co., Bronx, N. Y.

Fifth row: Iron City and Dutch Club, Pittsburgh Brewing Co., Pittsburgh, Penn./ all four Old German, The Queen City Brewing Co., Cumberland, Md./Rainier Club, Rainier Brewing Co., San Francisco, Calif./Select 20 Grand Ale, Red Top Brewing Co., Cincinnati, Ohio.

First row: Reisch Gold Top, Reisch Brewing Co., Springfield, Ill./Renner & Old German, The Renner Co., Youngstown, Ohio/Renner Old Oxford Ale, Renner Brewing Co., Youngstown, Ohio/ Souvenir, The Geo. J. Renner Brewing Co., Akron, Ohio/Sierra, Reno Brewing Co., Reno, Nev./Bruenig's, Rice Lake Brewing Co., Rice Lake, Wis.

Second row: Burgermeister, San Francisco Brewing Corp., San Francisco, Calif./Schmidt's Select, Jacob Schmidt Brewing Co., St. Paul, Minn./Schmidt's Ale, C. Schmidt & Sons, Philadelphia, Penn./Schmidt's quart, The Schmidt Brewing Co., Detroit,. Mich./next three Schmidt's, C. Schmidt's & Sons, Philadelphia, Penn.

Third row: both Schmidt's, C. Schmidt & Sons, Philadelphia, Penn./Heidel-Brau, Sioux City Brewing Co., Sioux City, Iowa/ Bon, Spearman Brewing Co., Pensacola, Fla./Cardinal, Standard Brewing Co. of Scranton, Scranton, Penn./both Sunshine, Sunshine Brewing Co., Reading, Penn.

Fourth row: Goldcrest 51, Tennessee Brewing Co., Memphis, Tenn./Champagne Velvet, Terre Haute Brewing Co., Terre Haute, Ind./Breda-Beer, "The Three Horse-Shoes" Brewery, Breda, Holland/ both Tube City, Tube City Brewery, McKeesport, Penn./Uchtorff, Uchtorff Brewing Co., Davenport, Iowa/Royal Bru, Union Brewing Co., New Castle, Penn./ Fort Schuyler, Utica Brewing Co., Utica, N. Y.

Fifth row: Fort Schuyler, Utica Brewing Co., Utica, N. Y./both Virginia's Famous and Olde Virginia, Virginia Brewing Co., Roanoke, Va./Wacker's Little Dutch, Wacker Brewing Co., Lancaster, Penn./ Gam, August Wagner Breweries, Columbus, Ohio/both Royal Amber and Wiedemann, The Geo. Wiedemann Brewing Co., Newport, Ky./Wooden Shoe, The Wooden Shoe Brewing Co., Minster, Ohio.

Signs / Bar Posters

Since United States breweries first started to establish and promote brand names in the last half of the 19th century, signs and bar posters have played an extremely important advertising role. In the three-year period from 1891-93 Pabst, for instance, spent $403,407 on its advertising, of which $164,253 (or 40%) was allocated to the category called "Signs and Views" (signs and pictures given to retailers).

While early signs were generally objects of beauty, such has not always been the case in the years since Repeal. As with so many other modern items, the emphasis seems to be more on quantity than on quality. Much of this change in emphasis is undoubtedly due to the Alcoholic Administration Act of 1935, which fixed at $10 the maximum value of all signs from a given brewery that could be displayed by a retailer at any one time.

Over the next few pages are grouped numerous signs and posters, arranged according to the type of material used in construction.

These are all glass signs, among the earliest type made. Early framed "reverse-painted" signs, such as the King's Bohemian above, command very high prices among collectors, especially if they are in very good or mint condition. Unfortunately, however, chipping or flaking often sets in (as with the King's and the smaller Wiener Beer) and mars the appearance of this type of sign.

METAL SIGNS

Along with glass, metal signs were the earliest type used to any real extent. Probably the oldest sign included on this page is the Mt. Pleasant Brewing Co.'s New England Pale Ale. Contrary to what you'd expect, however, this small brewery (10,000-barrel-per-year capacity) was not in New England, but instead was located in Mt. Pleasant, a village slightly northwest of Pittsburgh. As with Eberhardt and Ober, it was merged into the Pittsburgh Brewing Co. combine in 1899.

Cincinnati Burger Brau was a product of the Burger Brewing Co., a post-Prohibition Cincinnati brewery that went out of business in early 1973.

John Eichler, who founded the John Eichler Brewing Co. in the Morrisania section of the Bronx in 1865, had earlier been brewmaster for Jacob Ruppert's brewery. Eichler was born in Bavaria in 1829, and had learned the brewing trade in Baden and Berlin before coming to the United States in 1853. After many years as one of New York's most successful brewers, he died while on a trip back to Bavaria in 1892.

Notice that the Augustiner Bottled Beer moose head is the same one used by the Germania Brewing Co. on its tray, shown in the color pages. This very handsome scroll sign is somewhat puzzling in that, although it does say Reading, Penn., it does not say which Reading brewery it's from (and Reading has had quite a few).

METAL SIGNS with cardboard backing

This type of sign is generally newer than those of metal with no backing of cardboard. All of those shown here are post-Prohibition, with the possible exception of the "Ortlieb's In Bottles."

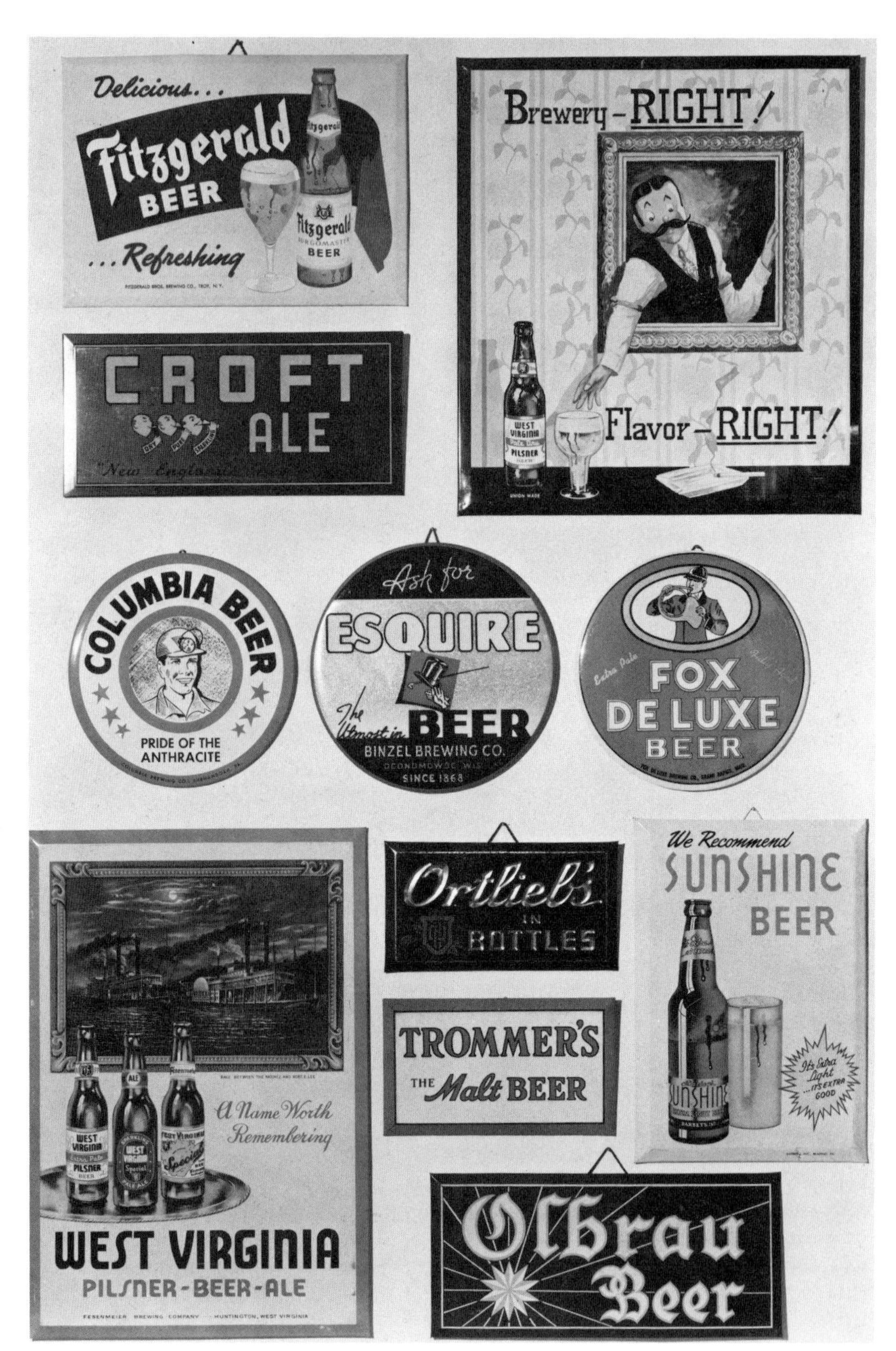

Without attempting to read the company name, can you guess where the Pride of the Anthracite, Columbia Beer, was brewed? Where else but northeastern Pennsylvania, America's anthracite (hard coal) region. The Columbia Brewing Co., organized in 1894, was right in the heart of Shenandoah, Penn., and remained in operation until fairly recently—1968.

I'm glad the Binzel Brewing Co. didn't name their beer after the Wisconsin town in which they were located; somehow Esquire seems easier to pronounce than Oconomowoc (especially after you've had a few beers). The Binzel Brewing Co. operated only five years, from 1937 to 1942. However, the company's namesake, Peter Binzel, operated a brewery in Oconomowoc as early as 1868.

CARDBOARD SIGNS

While cardboard signs, due to their comparatively low cost, were accepted early by American brewers, all of the signs shown here are post-Prohibition, except for Gund's Extra Pale.

Unfortunately, most of the brewery names do not show on these signs. So here they are, along with their dates of operation:

John Gund was a German who came to America in 1848, and worked in breweries at Galena, Ill., and Dubuque, Iowa, before settling in La Cross, Wis., in 1854. There he set up his own small brewery and operated it for four years before forming a partnership with Gottlieb Heileman. Their lager plant was known as The City Brewery, and all remained well until 1872, when Gund left to form his own John Gund Brewing Co. This was a very successful firm, increasing sales from 2,100 barrels in 1875, to 43,000 barrels in 1890, and to a very impressive 108,000 barrels in 1901. Success is fleeting, however, and Gund closed with the coming of Prohibition, never to reopen.

Success was much more lasting for Gund's partner, Gottlieb Heileman. When Gund left, Heileman took over operation of the brewery—and it has been operating ever since. Today the G. Heileman Brewing Co. brews in five plants and distributes its brands in 40 states and the District of Columbia. Among those brands, incidentally, are many famous names from once independent brewers—Wiedemann, Sterling, Blatz, (Jacob) Schmidt and Drewry's.

COMPOSITION SIGNS

It's hard to say what these signs are really made of. They're not wood or plastic, but some kind of a substantial, man-made substance; the term "composition" seems as good as any. Most of these signs are from the 1944-55 period, and are finished to look as if they're made of wood.

Who in the New York area can forget Piel's Less N.F.S. theme of the 1950's? N.F.S. meant Non-Fermented Sugar. What this means is anybody's guess. Notice the Cook's waiter (F. W. Cook Co., Evansville, Ind., 1853 to 1955) and then try to recall the famous Krueger "K man." There is quite a similarity between the two. Anteek was a product of the Chartiers Valley Brewery, Carnegie, Penn. (c. 1900 to 1953), while Old Nut Brown Ale was a seldom-promoted brand of the Duquesne Brewing Co. (1899 to 1972). If the two signs look alike, they should; Duquesne owned and operated Chartiers Valley during its entire post-Prohibition life.

MISCELLANEOUS SIGNS

An assortment of different type and different vintage signs. The Salt & Co.'s Burton Stout is a lovely nineteenth-century English sign, while the Robt. Smith's ½ and ½ and Reisch's Weiner Special are pre-Prohibition American signs. Reisch was an old-line Springfield, Ill., brewery that began business in 1849 and lasted to 1966.

Robert Smith was an Englishman who learned the art of brewing as an apprentice at the world famous Bass Brewery of Burton-on-Trent. His brewery, located at 38th Street and Girard Avenue in Philadelphia, was one of the largest and best known ale, porter and stout facilities in the country. Eventually absorbed by C. Schmidt & Sons, the Robert Smith formula is still used by Schmidt for their Tiger Brand Ale. The sign introducing "A New Summer Drink," incidentally, is a very heavy all enamel sign.

Why have we included the plastic, electric Potosi sign among collectibles? The answer is simple—even plastic and other newer items become meaningful when a brewery goes out of business. And, unfortunately, "Good Old Potosi," a small brewery in the Mississippi River town of Potosi, Wis., closed its doors forever in early 1973, after many years of brewing.

Old German, "The Beer Without a Peer," is still brewed in the hills of western Maryland by the Queen City Brewing Co. of Cumberland. If the Chester Brewing Co. (1898-1954) of Chester, Penn., were still in business, they probably would want to discontinue their "Silver Dime" brand, rather than to correct the name to "Copper Nickel Alloy Dime Beer."

Openers

Before the development of the crown top in 1892 there were, with one exception, probably very few bottle openers of any kind. That's because the vast majority of these closures were self-opening, i.e., they did not need any kind of tool or instrument to trigger them. The one exception to this, of course, was the corkscrew. Many ale and beer bottles did use corks throughout the late nineteenth century.

As the crown top steadily gained universal acceptance in the late 1890's and early 1900's, bottle openers, as we know them today, began to appear everywhere. It made no sense for breweries to install capping equipment and market their beer in crown bottles if their customers couldn't open the caps—so breweries distributed the necessary openers by the thousands.

As a result of this large scale distribution, early openers are still fairly easy to find today. Examine the small-item displays at almost any sizeable flea market and you'll almost certainly find some interesting examples.

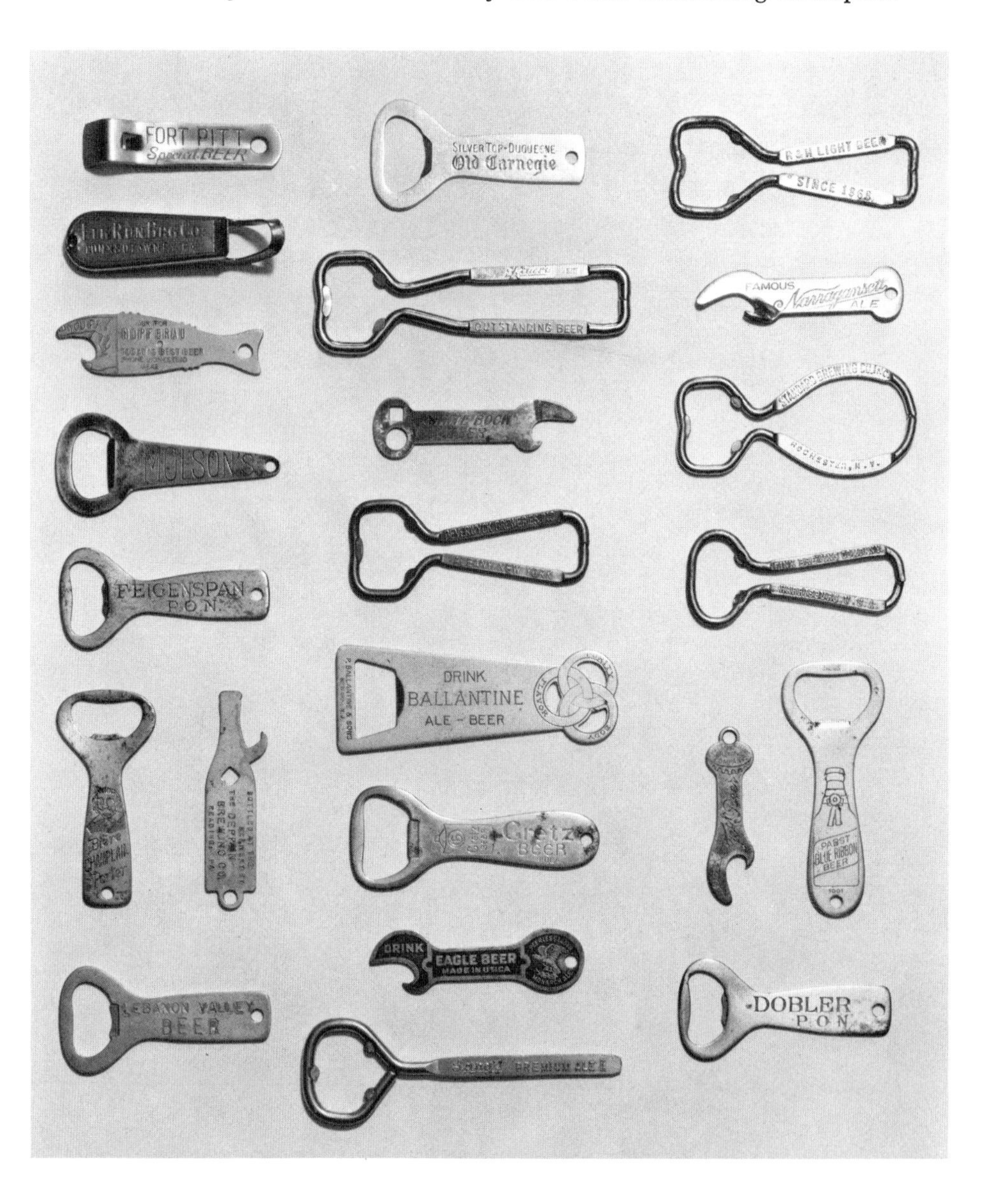

The most unusual item here is the Hopfbrau combination open-spinner (third from top, left row). Order a round, spin the "fish" and whomever it points to has to pay for the round. With a little luck you could drink free all night!

Some additional openers of note, including a fine beer corkscrew from Wacker & Birk, a large Chicago brewery founded in 1882.

The "Lady's Leg" Utica Club opener (top right) is also a penknife. By adding extra utility to an opener in this manner there was a better chance that it would be carried around in a prospective customer's pocket. And, of course, that was the whole idea, because with an opener at his fingertips a person could "tap" a bottle of brew at the slightest sign of thirst.

The little opener directly under the corkscrew is from the relatively short-lived Pacific Brewing & Malting Co., of Tacoma, Wash. Pacific opened up at the turn-of-the-century, and closed when Prohibition came early to Washington, in 1916.

While openers may seem like a colorless and mundane aspect of breweriana, remember that in today's "pop-top" and "twist-off" world, openers are rapidly becoming extinct.

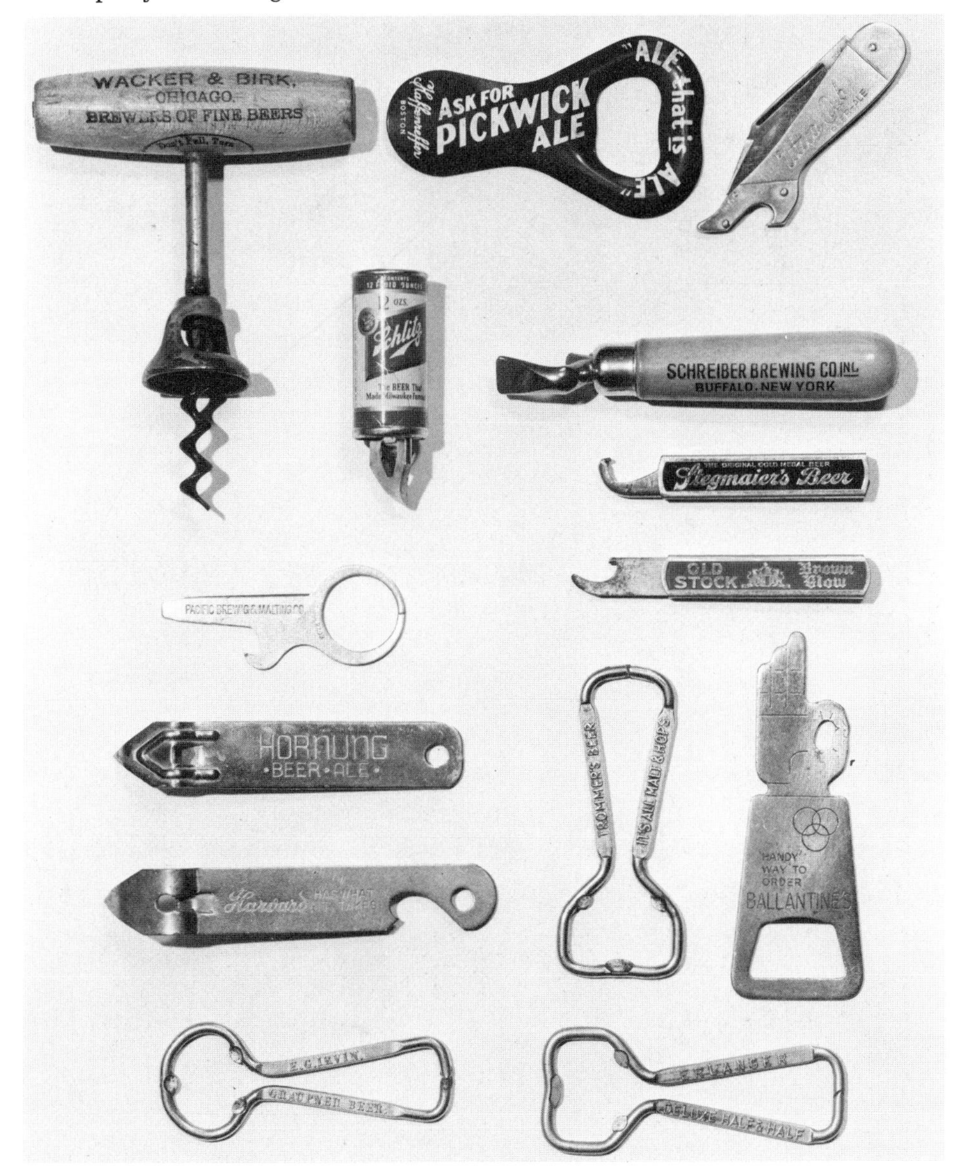

Also shown on this page are three can openers (Hornung, Harvard and the unusual Schlitz miniature can/opener). These are naturally more recent than most of the bottle openers, as the beer can was not introduced until 1935.

Brewery Plant Pictures

There is no form of architecture as consistently majestic and beautiful as brewery architecture. Admittedly, this is a somewhat biased view, but those who adopt breweriana collecting often develop the same love for these buildings. In fact, many brewery buffs feel it is almost as much fun to track down the old plants as it is to seek out trays, glasses or bottles at a flea market. With this feeling in mind (and heart) we've pictured numerous breweries of all ages, sizes and shapes over these next ten pages. As you look at them, remember that, with but a few exceptions, all the sparkling, frothy goodness and happiness that once poured forth from these buildings has long since ceased to flow.

THE OLD ECKERT BREWERY, READING, PENNSYLVANIA.
Founded by Henry Eckert, 1763. Now used as a residence.

Some very early breweries. Most of them look like houses, and, in fact, at least one of them originally was. From the Historical Society of Berks County, Pennsylvania, we find that "Henry Eckert built a two-story stone home on the east side of S. 4th St. between Franklin and Chestnut St. in 1763. He started a brewery, which he ran for a few years. In 1766 he sold it to John Spohn who sold it in 1775 to Jacob Bright. Mr. Bright operated the brewery until his death in 1813, after which his widow ran it for a short time. The building itself is no longer standing, being replaced by a barber shop, a church and several row homes."

BREWERY PREMISES GRANGER BREWING COMPANY, HUDSON, NEW YORK, 1859.

FIRST LAGER BEER BREWERY IN CHICAGO, 1847.
John A. Huck, Proprietor.

WIEDEMANN BREWERY, NEWPORT, KENTUCKY.

JOHN A. HUCK'S BREWERY, CHICAGO, 1871.

JOHN ORTH'S FIRST BREWERY, MINNEAPOLIS, MINNESOTA.

F. C. SELS BREWERY

Not all breweries have been torn down or turned into warehouses. There sits a very small brewery in Canyon City, Oregon, that has been lovingly restored by a dedicated group known as the Whiskey Gulch Gang.

The name Whiskey Gulch goes back to the days of gold panning. When the miners learned of an impending Indian attack, they hid their women and their whiskey in a gulch just south of Canyon City. The gulch is still known as Whiskey Gulch. F. C. Sels was founded in 1863. It burned down twice, in 1871 and 1898, but was rebuilt both times. Even though it was an extremely small facility, Sels is known to have done his own malting as well as brewing lager. The brewery went out of business circa 1912.

The "brewery" as it looked in 1971 when it was purchased by the Whiskey Gulch Gang.

Whiskey Gulch Gang, Inc.
Canyon City, OR 97820

April 13, 1973

Dear Mr. Anderson:

Just recieved your letter of inquiry concerning F. C. Sels Brewery in Canyon City. You're in luck, I can give you information about the old building.

The Whiskey Gulch Gang Inc. was formed in 1922 and has since on the first week-end of June each year held an old-time celebration to commerate the discovery of gold, June 8, 1862 in Canyon City.

Last year the gang bought the stone building that was built by F. C. Sels, from Tom Huntiongton of Canyon City. The building was being used at the time of purchase as a mechanic's garage. Before that the building served as a post-office and store.

The gang spent several thousand dollars and hundreds of man hours restoring the old building into an old time saloon. The interior is finished with a bar, back bar, old type wall paper, old photo's, western paintings and even the original swinging doors.

The building which is simply referred to as "Sels" is the frequented place during the '62 Days Celebration. Last year from Friday evening until Sunday evening, 53 kegs of beer were sold. Our local "Biltz" beer distributor said that per capita, Sels Saloon probably out sold any place in Oregon.

The building (outside appearance although repainted is practually identical to old photo's) is used throughout the year by the Whiskey Gulch gang for meetings, parties and often open for tourists to view.

If you're in this section of the country, stop in. We'll be glad to show you Sels.

Would also appreciate knowing when your book will be on sale

Sincerely

Dave Traylor
Whiskey Gulch Gang
President

Two years and numerous dollars and manhours later, Sels as it looks today after a magnificent restoration job.

Some of the "Gang" enjoying the fruits of their labor. During parties, it is not uncommon for over two hundred people to crowd themselves into the building to enjoy good times and good beer.

MIDWESTERN

Some nice Midwestern brewery plant scenes. The Cleveland & Sandusky Brewing Company resulted from the consolidation of nine northern Ohio breweries in 1898.

OLD GREINER BREWERY, ORIGIN OF THE MADISON BREWING COMPANY'S PLANT, MADISON, INDIANA.

PREMISES OF THE MATHIE BREWING COMPANY, WAUSAU, WISCONSIN.

GROUP OF BREWERIES OWNED BY THE CLEVELAND AND SANDUSKY BREWING COMPANY, CLEVELAND, OHIO.

FAR WESTERN

The Becker Brewing Co. opened for business in 1892 and remained in operation until the mid-1960's.

The Bellingham Bay Brewery is a rarity among the plants shown in this book in that it was built in the 20th century—1902, to be exact. The name "Whatcom" is perhaps more interesting than the brewery itself. While now a part of the City of Bellingham, it was originally an independent town, and a boom town at that. For a while during the Fraser River gold rush of 1858, Whatcom had 10,000 inhabitants, but it declined after that. The town's unusual name comes from its being named after an Indian chief, Chief Whatcom of the Nooksacks. The word is said to mean "noisy water."

The Buffalo Brewing Company's beautiful facility was built in 1889 and produced many thousands of barrels of good Buffalo beer before it was closed in the late 1940's. Sometime between 1948 and 1951 this magnificent structure was torn down to make way for the Sacramento Bee newspaper building, which presently occupies the brewery's old 21st and Q St. location.

The Zang Brewery almost predates Denver. It was originally founded in 1859, only one year after Denver was first settled, as the Rocky Mountain Brewery. Philip Zang, an experienced brewer born in Bavaria, became owner of the brewery in 1870 and started to enlarge the facility, making it within a very few years far and away the largest brewery in Colorado.

PREMISES OF THE BECKER BREWING AND MALTING COMPANY, OGDEN, UTAH.

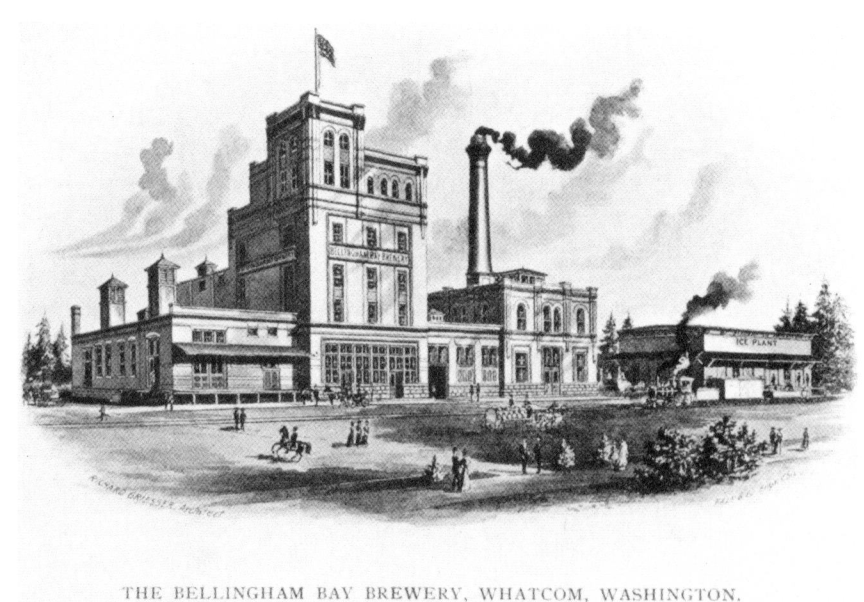

THE BELLINGHAM BAY BREWERY, WHATCOM, WASHINGTON.

After Prohibition, while there was considerable talk about building a new brewery to adjoin the original structure at 7th and Water Streets, I don't believe the company was ever reopened. In any event, the original buildings burned Nov. 2, 1971. A gas station now adorns much of what was once the brewery's location.

PREMISES OF THE PHILIP ZANG BREWING COMPANY, DENVER, COLORADO.

SOUTHERN

Because the warm climate was not conducive to the lagering of beer, there were few breweries founded in the deep South in the nineteenth century. Here and there, however, were exceptions to this rule—as with the three breweries pictured here.

Mobile, about as far south as you can go, had its first brewery of any real consequence when the Mobile Brewing Co., located at Adams & Water Sts., opened for business in April of 1892. This Gulf of Mexico port city had but one other pre-Prohibition brewery—the very short-lived Bienville Brewery on St. Joseph & Broodgood Sts.

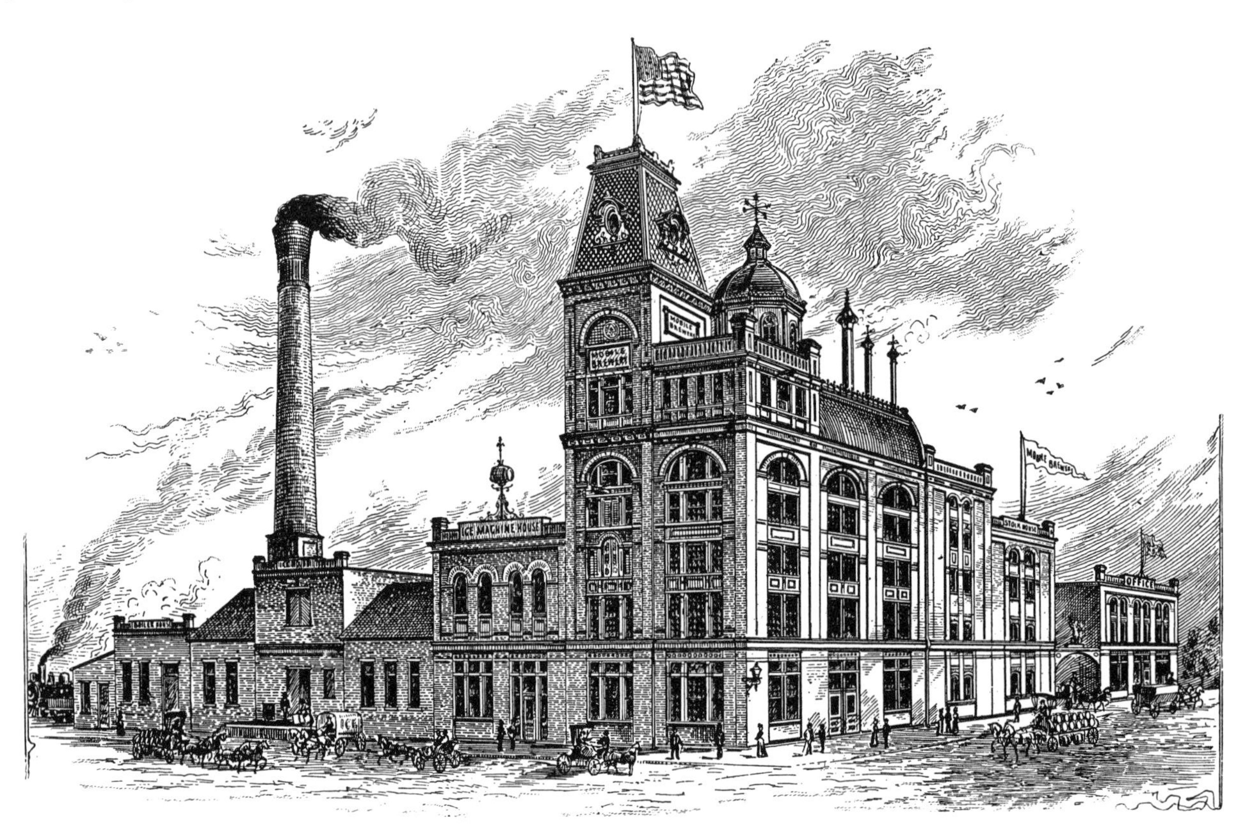

PREMISES OF THE MOBILE BREWING COMPANY, MOBILE, ALABAMA.

The American Brewing Association of Houston was founded in 1895, and was backed by none other than Adolphus Busch. Busch, as a matter of fact, invested heavily in a number of southern breweries in the last quarter of the nineteenth century. These included the Lone Star Brewing Co. and the Alamo Brewery, both of San Antonio; Shreveport Ice & Brewing Co. in Shreveport, La.; the Texas Brewing Co. of Fort Worth; as well as American.

The Tennessee Brewing Co. was originally organized in 1885, but the plant shown here was built in 1892, and produced its last brew 62 years later, in 1954. Still standing, the building today is used by a scrap metal dealer, and was written up in a recent (April 2, 1973) article in the *Memphis Commercial Appeal*. One paragraph in this article really stands out in my mind because in many respects it applies to virtually all old breweries, not just the Tennessee Brewing Company: "The building is a classic. It has the ornate arches, the wrought iron, the charm of another day. It is hard to imagine that it was built as an industrial site and that 12 million barrels of Goldcrest 51 beer and Columbian Ale were produced there."

Not much is known about the Rock Brewery, except that it was founded in New Haven by George Basserman prior to 1878, and went out of business in the early 1890's. This is an actual photograph found at a New Haven antiques show, and it indicates that posing for the brewery's formal portrait was a grand and glorious affair. Centlivre is shown here in an early 1900's lithograph. Charles L. Centlivre, the company's founder, was a Frenchman, born in Alsace in 1827. As a young man he came to this country and at first practiced his original trade of cooper. In 1862 he settled in Fort Wayne and, along with his father and brother, built a crude brewery with his own hands, the beginnings of the plant pictured here. As mentioned, virtually all of the breweries shown in this section have long since ceased to be operative. This is an exception. On July 1, 1961 the employees of the Centlivre Brewing Co. jointly purchased the company, and have continued to operate it, as the Old Crown Brewing Corp. Old Crown is justly proud of the fact that it is the only 100% employee-owned brewery in the U. S., and possibly the world.

Two classic brewery shots, contrasting a large plant with a fairly small one.

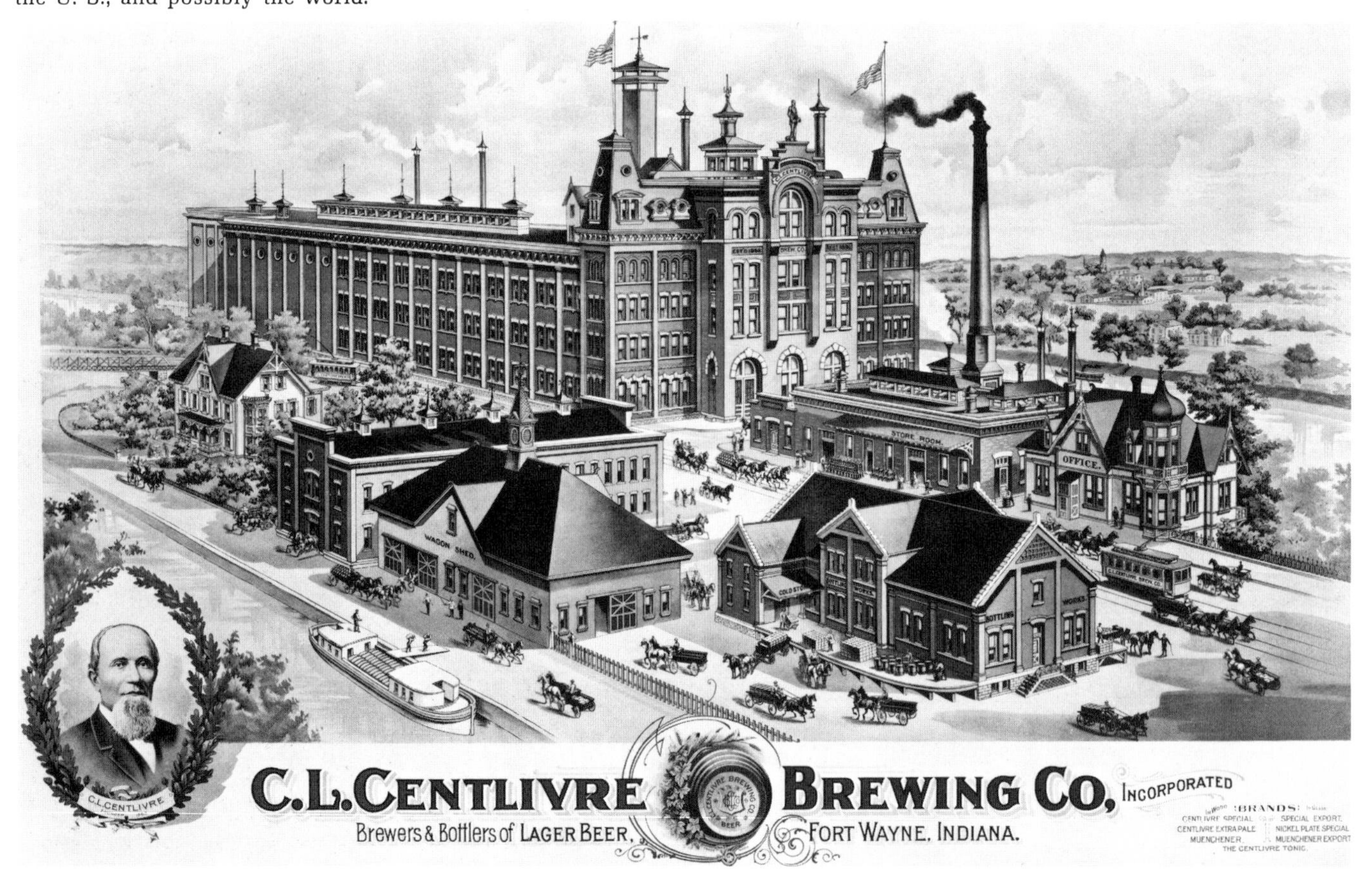

Here are two more interesting brewery scenes—from entirely different sources. The John Taylor scene is from the 1870 New York Business Directory, while the Fred Krug Brewing Co., of Omaha, Nebraska, issued this plate to commemorate their 50th Anniversary in 1909.

This plant, a very substantial one for its day, was built in 1851 on a man-made plot built into the Hudson River.

In the mid-nineteeth century Albany was about as famous for its ale as Milwaukee was later to become for its lager. And Taylor, who began brewing in Albany in 1822, was perhaps the most famous of all the Albany ale brewers.

The brewery, as happened to so many other ale companies, went out of business in 1909. Five years later, on May 22, 1914, this once very bustling facility was destroyed by fire.

Shown are Krug's original 1859 brewery and his much more substantial plant 50 years later. The original structure was at 1013-15 Farnam St. and he outgrew it in eight years, moving to a new plant at Jackson and 10th Streets in 1867.

Because both his sales and Omaha were booming, Krug, in 1892, moved again, to another brand new plant, at 3302 S. 25th St. This is the plant shown in the middle of the plate—and it's still in operation today as a Falstaff branch brewery. Falstaff leased the plant in 1935, and purchased it two years later.

It's a shame that this A. Finck & Son's poster, almost Currier and Ives in its appearance, was "repaired" by an earlier owner. Instead of taping it on the back, he taped it on the front, marring an otherwise fabulous early New York City "Bier Brewery" scene. Where once this brewery stood is now—you guessed it—a parking lot.

Bottle Enclosures
Caps

What boy (or girl?) has not saved bottlecaps. My favorites here are "It's My Beer" and the two Katzenjammer Kids "Comicaps." These were used by the Hittleman-Goldenrod Brewery of Brooklyn, N. Y., right after Repeal.

Both "The Captain" and "Fritz" caps shown carry a 1935 copyright date. Unfortunately, while many of these caps show the name of the brewery as well as the brand name, many do not—making it pretty much of a guessing game as to who the brewer was.

As discussed in the section devoted to bottles, up until the crown top's adoption in the 1890's, there were literally hundreds and hundreds of stoppers and enclosures patented by inventors, and tried by bottlers. Quite a few revolved around the use of a porcelain stopper. Because these generally have the name of the brewery baked into them, they are now considered collectible—predominantly by bottle collectors.

Here's how the porcelain stoppers look when closed. Snap the metal bail down and the stopper pops up, ready to pour.

Some loose porcelain stoppers. Long after the metal bail has rusted away these porcelain tops will still be as good as new. Also shown is a Hamm's cork stopper (probably meant to be used once the bottle was opened) and two unusual enclosures from Charles Weinacht, a Hoboken, N. J., liquor dealer.

Labels

As discussed in the section on bottles, labeled bottles are not that avidly sought by breweriana collectors. This is not the case, however, with the labels themselves. In fact, label collecting may well be the single largest breweriana field. Certainly labels are the most universally collected beer item, and it is relatively easy to develop trading relationships with people from distant and often intriguing foreign countries. Naturally, labels can be obtained by soaking them off the bottles. Obtaining them fresh and crisp from the brewery itself is far more preferable. The collector, however, may find that many brewers have understandably curtailed their supply in the past few years due to a more-than-they-can-handle increase in volume.

Early labeled bottles were truly a thing of beauty. It's hard to believe these, especially the Beadleston & Woertz, contained beer, and not champagne.

EARLY EASTERN LABELS

Private label beer has meant different things over the years. Now pretty much relegated to supermarket and liquor store discount brands, private label beer was once the opposite—the top-of-the-brewery's line. Two examples of this are represented by the Betz and Geo. Ehret's Special Brew labels. Ehret's is especially meaningful to New Yorkers as "Meet Me at the Astor" was a favorite slogan for generations, until this beautiful old Times Square hotel was torn down in 1966.

One of the most puzzling of all breweriana questions is—how extensively were labels used by pre-Prohibition breweries? Phrased differently—how many breweries used only embossed bottles; how many used only labeled bottles; and how many used a combination of the two? Judging from this page of randomly selected pre-Prohibition New York City and Brooklyn labels, the answer may well be that a goodly percentage used a combination.

Embossed bottles from Congress, Obermeyer & Liebmann, Clausen-Flanagan, Ferd. Neumer and John F. Betz's Manhattan Brewery are all shown in this book. Yet here are labels from the same firms. It could be, of course, that the labels came about because of a switchover from embossing to labeling. It could also well be that labels were used for special runs—to "dress-up" beer bottles for export, for "special brews" or for private label usage. In any case, it's an interesting subject, and one that needs more research.

WISCONSIN LABELS

Wisconsin, even since Repeal, has had many, many breweries, ranging in size from the giants of Milwaukee to almost "home-brew" operations producing but a few thousand barrels a year. Represented on this and the next page are just a few of these many companies, including for each the year operations ceased.

Top row: Blatz Brewing Co., Milwaukee, 1958 (although later operated by Pabst through 1969)/Calumet Brewing Co., Chilton, 1942/Capitol Brewing Co., Milwaukee, 1948.

Second row: Blatz Brewing Co., Milwaukee, 1958/The Cassville Brewing Company, Cassville, 1938/Fauerbach Brewing Co., Madison, 1966.

Third row: Harold C. Johnson Brewing Co., Lomira, 1955/William G. Jung Brewing Co., Random Lake, 1958.

Bottom row: Mathie-Ruder Brewing Co., Wausau, 1955/Medford Brewing Co., Medford, 1949/Milwaukee Germantown Brewing Co., Germantown, 1941/Mineral Spring Brewing Co., Mineral Point, 1966.

Top row: The North Lake Brewery, North Lake, date unknown/Oshkosh Brewing Co., Oshkosh, 1971/Plymouth Brewing Co., Plymouth, 1937.

Second row: Potosi Brewing Co., Potosi, 1973/Rhinelander Brewing Co., Rhinelander, 1967/Stevens Point Brewery, Stevens Point, still in business.

Third row: Walter Brewing Co., Eau Claire, still in business/Walter Bros. Brewing Co., Menasha, 1956.

Fourth row: Wausau Brewing Co., Wausau, 1961/Weber Waukesha Brewing Co. (Kingsbury), Sheboygan, still in business/Walter Brewing Co., Eau Claire, still in business.

Bottom row: Louis Ziegler Brewing Co.. Beaver Dam, 1954.

LABELS, A to Z

Here's a brief look at some additional labels, arranged A-Z by brewery.

Top row: Ambassador Brewing Co., Los Angeles, Calif., 1934-1940/American Brewing Co., Baltimore, Md., 1934-1973/Anthracite Brewing Co., Mt. Carmel, Penn., 1897-Prohibition.

Second row: Bay City Brewing Co., Bay City, Mich., 1884-1945/Deppen Manufacturing Co., Reading, Penn., 1879-1937/Dick Bros. Brewing Co., Quincy, Ill., 1857-1952.

Third row: Dubuque Star Brewing Co., Dubuque, Iowa, 1898-1971 (still in business as Joseph S. Pickett & Sons)/Fesenmeier Brewing Co., Huntington, W. Va., 1893-1968/Fischbach Brewing Co., St. Charles, Mo., c. 1907-1965.

Fourth row: Ernst Fleckenstein Beverage Co., Faribault, Minn., early 1870's-1964/Geyer Bros., Frankenmuth, Mich., 1862, still in business/Robert H. Graupner, Harrisburg, Penn., 1895-1951 (known as the Harrisburg Consumers Brewing & Bottling Co., until 1903).

Bottom row: A. Haas Brewing Co., Houghton, Mich., 1859-1942 (brewery moved to Hancock, Mich., remaining in operation there until 1953)/Home Brewing Co., Richmond, Va., 1891-1969/Kamm & Schellinger Co., Mishawaka, Ind., 1879-1953.

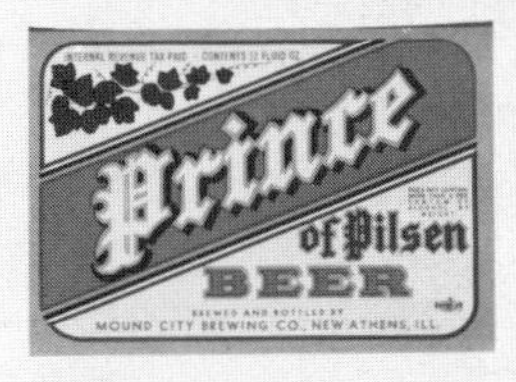

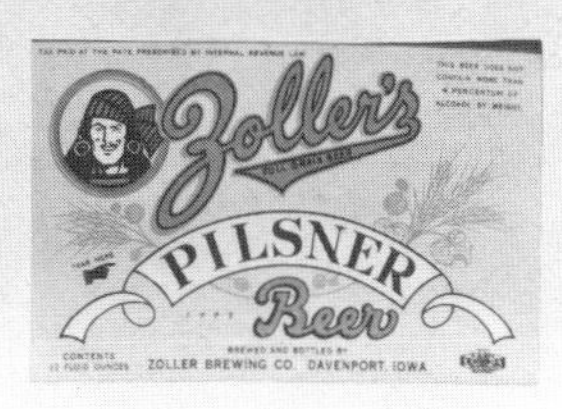

Top row: Kessler Brewing Co., Helena, Mont., 1865-1958/Lebanon Valley Brewing Co., Lebanon Valley, Penn., 1934-1959/ Menominee-Marinette Brewing Co., Menominee, Mich., 1933-1961.

Second row: Minneapolis Brewing Co., Minneapolis, Minn., 1890–still in business (name changed to Grain Belt Breweries in 1967)/Mound City Brewing Co., New Athens, Ill., 1933-1951/Penn Brewing Co., Steelton, Penn., opened and closed in 1934.

Third row: Peoples Brewing Co., Duluth, Minn., c. 1907-1957/Peter Bub Brewery, Winona, Minn., 1870-1969/Reisch Brewing Co., Springfield, Ill., 1849-1966.

Fourth row: Reno Brewing Co., Reno, Nev., c. 1900-1958/The Stanton Brewery, Troy, N. Y., c. 1880-1951; Sicks' Missoula Brewing Company, Missoula, Mont., 1933-1964 (known as Sicks', 1946 to 1949).

Fifth row: Reno Brewing Co., Reno, Nev., c. 1900-1958/Sprenger Brewing Co., Lancaster, Pa., 1895-1951.

Bottom row: Star-Union Products Co., Peru, Ill., 1856-1966/Zoller Brewing Co., Davenport, Iowa, 1892-1944 (in 1944 changed name to Blackhawk Brewing Co.; in 1953 name changed again to Uchtorff Brewing Co., and remained in business until 1956).

Holiday Items

Christmas and other holidays have often been the occasion for special holiday advertising and packaging of all kinds of products—and beer is no exception.

Many companies, through the years, have made special Holiday or Christmas brews.

A pre-Prohibition trolley-car poster from Bartels. Over a half-century later, I can't think of a nicer way to say "Merry Christmas!"

Cards

Included here are a number of different types of cards. The Hubert Fischer item is an early (there's a faint date of June 24, 1890) form of postcard used long before the impersonal 1960's and '70's.

The North Western "mug" was obviously given as a souvenir to people who visited the brewery's exhibit at the 1893 World's Fair, but it is difficult to determine the exact use of the American Brewing Co. "twin bottle" card. In any case, the front sides of both are pictured in full color on the color pages in this book.

The two table scenes are postcards from "Sieben's Bier Stube," 1466 Larrabee St., adjoining Sieben's Brewery in Chicago. This was a small brewery, founded by Michael Sieben in 1876, and continued in operation until 1967. Amazingly, the brewery put out its Real Lager only in kegs and quarts.

Hartford, ______

Dear Sir:

Please ship all empties in your possession as soon as possible and oblige,

Yours respectfully,

HUBERT FISCHER

Coasters

Coasters, something we generally take for granted, have actually had a long and illustrious history. Originally made of pottery or porcelain, the wood-pulp coaster as we know it today came into usage in 1892 when Robert Smith of Dresden, Germany, developed and patented it. While other materials (glass, metal, etc.) are still used for coasters, wood pulp is so economical that it's the type used almost exclusively by breweries.

Because they are so economical, coasters are often designed for special events, used for an appropriate short period of time, and then discontinued. Two examples of this practice are pictured here—the Ballantine Inn 1939 World's Fair coaster, and Rheingold's more recent sponsorship of the World Champion New York Mets (1969). The Ballantine coaster, given as a Fair souvenir, was also a postcard. On the reverse side, in addition to a limerick praising Ballantine, there is room for a stamp and the recipient's name and address.

Coaster designs are generally changed with regularity, so the collecting of sets has become a legitimate and lively interest. The Binding Brewery of West Germany has a set going that now exceeds 300—one side is always the same, but the second side features constantly changing cartoons. Cartoons have been popular in the United States as well—as witnessed here by samples of two different sets of Piel's.

Related Industries Objects

While the purpose of *The Beer Book* is to focus attention on, and interest in, the brewing industry and its fascinating collectibles, the trades and industries that serviced and supported brewing are almost as interesting—and should not be overlooked by any breweriana fan or collector.

Several sample bock beer posters or "hangers." These were produced by lithographers to show their "stock designs," demonstrate their quality, and highlight their reasonable price. And, in retrospect, prices were extremely reasonable; just try today to get 100 of any piece of art printed in five colors for $4.00!

Several 1890's related industry ads aimed at St. Louis brewers.
Stecher Cooperage was in business from 1890 until 1919; Seibel-Suessdorf from 1885 to 1959; Kupferle Bros. Mfg. Co. from 1877 to 1922, and A. Ruemmeli-Hacker for a very short period of time around 1896.

If you were a Philadelphia brewer around the turn-of-the-century these were some of the companies with which you'd undoubtedly be doing business.

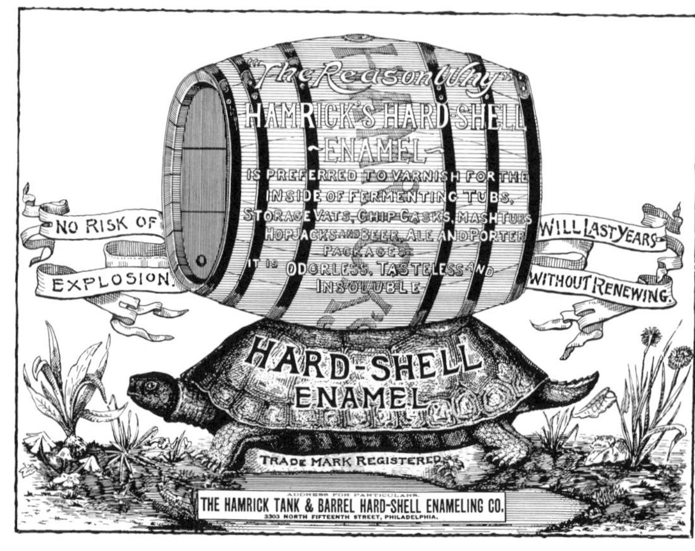

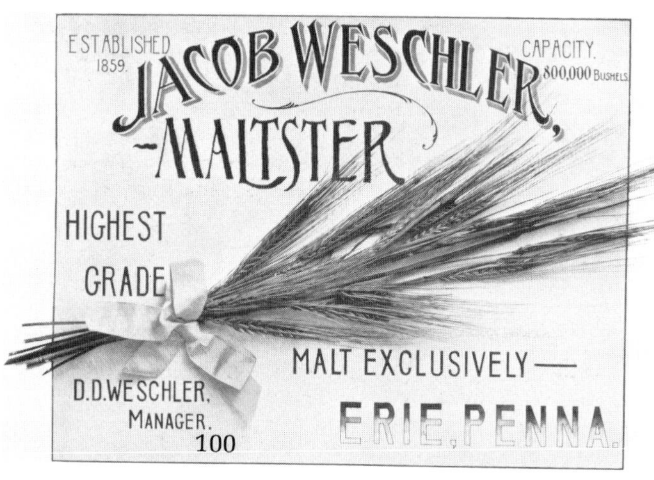

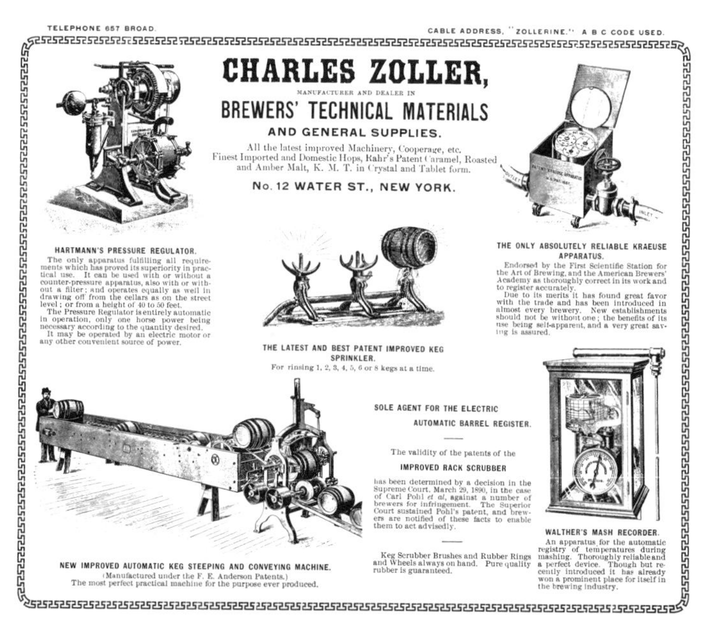

Miniature Bottles

These little bottles, 3″ x 4½″ in height, are very collectible, and authentic in that they are exact reproductions of the brewery's actual bottle, much like liquor "nips." There, however, the comparison ends. While "nips" (the miniature liquor bottles used on airplanes and trains) contain liquor, post-Prohibition miniatures rarely contained beer. They were either empty, for use as salt and pepper shakers, or were filled with "frothy water."

Distributed almost entirely from 1934 to 1942, the mini-beers were used for advertising promotion, brewery tour giveaways and, on occasion, as commemorative or World's Fair souvenirs. Some miniatures were known to have been produced prior to Prohibition. These, of course, were more apt to be embossed, as well as labeled, and generally did hold beer.

Some examples of miniature beers, all post-Prohibition. Twelve of the 23 bottles shown here are meant to be used as salt and pepper shakers, including the Coors on the bottom row. This is an unusual item in that it is ceramic, with the label baked in.

The Koppitz Silver Star and the Acme next to it (both top row, left) are not actually bottles. They are made of wood and are really openers. However, they fit nicely in a collection of miniatures. This is also true of the small Budweiser bottle (shown front and back) on the bottom row. It's made of metal and opens up to become a cigarette lighter.

Miscellaneous

Here's a little bit of everything, from a Kaier pot-holder to a Lauer Brewing Co. mirror.

The strange looking Haberle item in the right forefront is a match holder. Made of very heavy pottery, the matches (wooden) stood up in the top and could be lighted by striking against the rough surface on the side.

Even more unusual is the tinware item on the left. It's a three-tier lunch box, embossed, The Highland Brewing Co., Springfield (Mass.). Highland opened up in 1894 and was merged into the Springfield Brewing Co. in 1899, so this, the fanciest lunch box you'll ever see, was from the "Gay '90's." The Barmann item is made of very sturdy wood, and is a stirrer-holder. It's from after Prohibition when Barmann had rights to the Evans Ale name.

The hand reaching out at you was meant to hold a display bottle of Tad, short for Tadcaster, a brew produced in Worcester, Mass., by the Brockert Brewing Co. (1936-1945) and its successor, the Worcester Brewing Co. (1946-1962). Ducking away from "the hand" is Fred Krug, founder of the company honored by this 50th anniversary plate.

What better way to cool off on a hot summer night than with a cold beer and a large fan!

Illustrated are a number of miniature miscellaneous items. The "Compli-

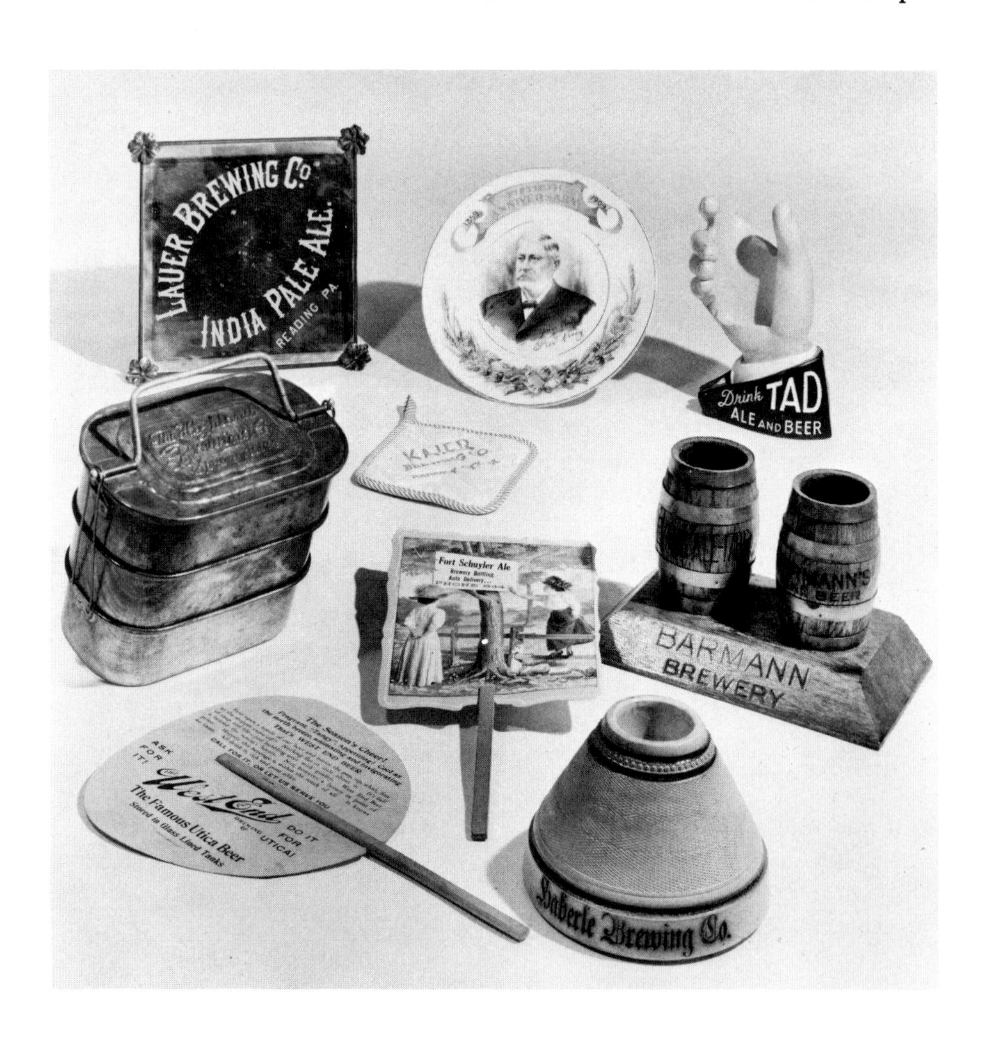

ments of Leibinger & Oehm" metal piece was a menu-holder. It slides open and inside there's a small piece of paper on which the tavern owner could list his sandwiches or special of the day.

Crockery City Ice & Products Co. was an East Liverpool, Ohio, brewery founded right around the turn-of-the-century. In 1946 it was sold to the Webb Corporation, who operated the plant until 1952. During that time they issued the White Label coin shown in the upper right. It's actually, on the reverse, a spinner to determine who buys the next round. Spin the coin and whomever the arrow points to is the lucky "winner." The coin on the lower right really was a coin, good for one Loyalhanna Beer at a participating neighborhood bar (which was most likely to be found in Latrobe, Penn., the location of the Loyalhanna Brewing Co.).

The Bowler Bros. item is a pre-Prohibition match safe, so named because it kept wooden matches safe and dry.

The Louis Bergdoll object featuring the luxuriantly coifed lady is a fine

Louis Bergdoll mirror, courtesy of America Hurrah Antiques, N. Y. C.

example of the pocket or purse mirror given away by breweries and other consumer goods companies in the turn-of-the-century era.

The Third Avenue El medal was issued by Ruppert, a New York Third Avenue brewery, to mark the passing of the elevated railway in 1955. Ruppert's own demolition came but fourteen years later, in 1969.

By saving pennies in their penny-holder, eventually you would have enough money to buy a Daeufer's (Daeufer-Lieberman Brewing Co., Allentown, Penn.).

Last is the pin shown in the bottom left. From a "beer blast" of long ago, it reads "Drink Wiedenmayer's Pure Beer at the 21st National Saengerfest, held in Newark, N. J., June 30 to July 6, 1906, and you will drink it ever afterwards."

This page further serves to show just how wide the range of breweriana collectibles is. Match books have been collected by the millions, and here are some that advertise beer. As you can tell from the label designs and especially the 10¢ price for Atlantic, these are fairly old—probably late 1930's. It's appropriate that Pabst is included here because it was Pabst, in 1899, that obtained exclusive rights for beer advertising on match books. They received these rights by giving Diamond Match, the company that developed the match book as we now know it, a 10,000,000-book order.

The letterheads are self-explanatory, and the Louis Bergdoll item is a small memo pad. The two coupons above the pad, however, are worthy of note. It seems that, in New York City anyway, there was a time when these coupons were used for returnable bottles. The face of the coupon reads—"The person to whom this ticket is issued has deposited with the undersigned *three cents* as security for the return of three bottles stamped with one of the names printed on the reverse of this ticket, which said bottles the said person hereby agrees to return when empty.

"Upon the return of said bottles and the surrender of this ticket, three cents will be refunded." These coupons came in three- and six-cent denominations. Getting their returnables properly returned always was a problem for breweries!

While virtually all United States breweries began as one-man, family or partnership enterprises, many later "went public." Shown here are stock certificates from four such companies.

Also illustrated are an 1890 Wm. Hull & Son invoice, and two brewing company checks. Hull, who brewed *only* ale and porter at the time of the invoice, eventually added lager to his line. The company is still in business today, Connecticut's only brewery. They are no longer on Whiting Street, however. After Repeal, Hull reopened at 820 Congress Avenue in New Haven—the former home of the pre-Prohibition Philip Fresenius' Sons Brewing Co.

These are known as beer combs or foam scrapers. The last term is the more accurate, as they were used to clear the foam off the top of a mug or pitcher of beer. As yet another form of advertising, the breweries would distribute them free to taprooms. Rumor has it that foam scrapers were discontinued for sanitary reasons.

Four uniform patches. Nothing very exciting about a piece of cloth sewn on a deliveryman's uniform, but they are usually quite colorful, and they are also break-proof and rust-proof!

Stoney's, a product of the Jones Brewing Co., of Smithton, Penn., has an interesting story behind its name. W. B. Jones, President of the company, writes "The name 'Stoney' was a nickname of my Grandfather, who was the founder of our corporation. The original name of our beer was Eureka Gold Crown Beer, but most people in this area simply asked for 'Stoney's' Beer, meaning the beer brewed by my Grandfather. After Prohibition, when we reorganized, we simply flowed with the tide, and renamed the beer Stoney's Beer."

REPRESENTATIVE AMERICAN BREWERIES

Two early large pottery or earthenware jugs. The Tennants', which even has a pour spout, is from England, while Smith Bros. was a New Bedford, Mass., brewery founded in 1906. Since Smith Bros. were also a wholesale liquor dealer, this five-gallon jug may well have held whiskey and not beer.

Introduction

Most everyone, from the most devoted beer connoisseur to the occasional-bottle-of-beer-on-a-hot-summer-day person, has his or her favorite brand of brew. Many of these brands, especially the breweries that brew (or brewed) them, have interesting histories. We've included short biographical sketches of numerous United States breweries, plus "family portraits" of some of their most interesting advertising and packaging.

Our goal has been to make this section representative of the American brewing industry's past century. Therefore, while we have included several of the largest breweries in the country, we're also proud to include several of the smallest. The oldest brewery in the country and the oldest brewery west of the Mississippi are both included. And there are many companies that, sadly, are no longer in existence—for the one fact that has characterized, and continues to characterize, the United States brewing industry is the steady decrease in number of operating units.

ANCHOR STEAM BEER BREWING COMPANY

541 8th STREET
SAN FRANCISCO, CALIFORNIA

Anchor Steam is today the only descendant of a once hardy American innovation in brewing—steam beer.

As explained on the brand's label, "Steam Beer, America's only native beer, originated during the California Gold Rush. In those days ice came around the Horn from Boston and was too costly for brewing. But in San Francisco, famous for its cool climate, a special formula evolved for brewing without ice. The new beer took its name from its characteristic creamy 'head of steam.' "

In the late 19th century, there were dozens of steam beer breweries on the West Coast, mainly in Northern California. San Francisco alone had 27. As lager was steadily introduced in the years following the turn-of-the-century, however, steam beer sales showed a sharp decline. By the time of Prohibition, only seven steam breweries were left, and since Repeal there has been only one—Anchor Steam. Correctly calling itself the only steam beer brewery in the world and the smallest brewery in America, Anchor was solely a draught operation until the late 1960's when it also began to bottle some of its frothy and distinctive brew. The company's present payroll consists of five people, headed by owner and brewmaster Fritz Maytag, great-grandson of the washing machine maker.

EXCELSIOR STEAM BREWERY
Imperial XXX Ales and Porter of the best quality, in Hogsheads, Barrels and Half-Barrels, constantly on hand and warranted.
CHARLES MURPHY, Jr.
Nos. 15, 17 & 19 Dutchess Avenue, Poughkeepsie, N. Y.

Although steam beer as Anchor Steam brews it was purely a West Coast phenomenon, the East had its own "steam beer." Loerzel Brothers was listed as a Saugerties, N. Y., brewery in the early 1890's; the Alexander Steam Bottling Works in Kinderhook, N. Y., produced (or at least bottled) Alexander's Excelsior Beer; and both M. J. Doran and Charles Murphy, Jr., made heavy reference in their advertising to steam. The use of the term "steam" by these breweries, however, most certainly refers to the Pasteurizing process. Known as "steaming" for quite a number of years before Pasteur was properly honored, this was a process used by virtually all good bottlers from the 1870's on. These beers, therefore, were actually "steamed" rather than "steam."

ANHEUSER-BUSCH, INCORPORATED

721 PESTALOZZI STREET
ST. LOUIS, MISSOURI

"A hole in the ground, supported by neither brick nor stone wall, being the cellar, with a board shanty over it, for the brew house." Such were the very humble beginnings of what is now the world's largest brewing company. This passage, from *One Hundred Years of Brewing,* describes the primitive brewery constructed by Georg Schneider in 1852. In the early 1850's Schneider was succeeded by Messrs. Hammer and Urban. They also failed, and in 1860 one of Hammer and Urban's principal financial backers, Eberhard Anheuser, purchased the facility. The successful growth of the company dates from that year, although it is Adolphus Busch who is generally given most of the credit for making Anheuser-Busch the giant it is today.

Busch was born in 1839 in Kastel, Germany, the son of a well-to-do landowner. Schooled in Brussels, he learned English and French in addition to his native German. In 1857, Busch emigrated to America and St. Louis. There he successfully involved himself in the brewer's supply business. In 1861 he married one of Eberhard Anheuser's daughters, Lilly, and later was made a partner in the brewery. From 1860 to 1875 the firm was called E. Anheuser & Co., and then was changed to E. Anheuser Co.'s Brewing Association. In 1879 it became Anheuser-Busch Brewing Association.

Ironically, it was also in the 1870's, specifically 1876, that Budweiser, the brand that was to make Anheuser-Busch famous (and vice versa), was introduced. At the time Anheuser-Busch was already marketing many other brands of beer, but Budweiser soon passed them all in sales. The formula for "The King of Beers" was developed by Adolphus Busch and a close friend, Carl Conrad. In fact it was Conrad who first registered the Budweiser trademark, in 1878, as his own private label. So, while Anheuser-Busch brewed the beer, it was Conrad who bottled it and sold it under the Budweiser name. It was not until 1891 that Anheuser-Busch obtained the rights to the Budweiser name.

Anheuser-Busch passed the 1,000,000-barrels-sold mark in 1901, the second American brewery to do so. The first had been Pabst in 1893. Shortly after the turn of the century, the Jos. Schlitz Brewing Company also joined the ranks of the "1,000,000 Club." Now, almost three-quarters of a century later, these three are still battling it out, only with the tables somewhat turned. In 1972 Anheuser-Busch ranked first with estimated total sales of 26,522,000 barrels; Schlitz, second with estimated total sales of 18,906,000 barrels; and Pabst, third, with an estimated total sales of 12,600,000 barrels.

A fine assortment of Anheuser-Busch breweriana, including two pieces of sheet music that heralded Budweiser and "Anheuser-Bush."

The bottle line-up includes a very early E. Anheuser & Co. and a C. Conrad & Co. original Budweiser, as well as several early branch bottles. Notice the very prominent "A and Eagle" trademark, originated by Eberhard Anheuser in 1872 and still in use today.

The top can is known as the "Gold Claw" design and was the type used to introduce Budweiser in cans in August, 1936.

Pictured on the large oval tray, in beautiful detail, is the St. Louis plant as it looked shortly after the turn-of-the-century. Because the Adolph Coors facility in Golden, Colorado has recently added increased capacity, Anheuser-Busch can no longer call its St. Louis plant "the largest brewery in the World." The company can, of course, continue to state that they are "the World's largest brewer."

Michelob, advertised here on a 1933 sign, was introduced in 1896.

An 1895 ad for White Label Exquisite. Anheuser-Busch dropped the "Brewing Ass'n." in 1919, in favor of its present-day Anheuser-Busch, Inc.

Prior to Prohibition, there were several different Budweiser girls. Shown here, on the cover of a 1909 booklet put out by the brewery, is the "Dutch Girl."

Even though it was founded in 1897, five years after the introduction of the crown top, DuBois still used blob-top bottles in its early days.

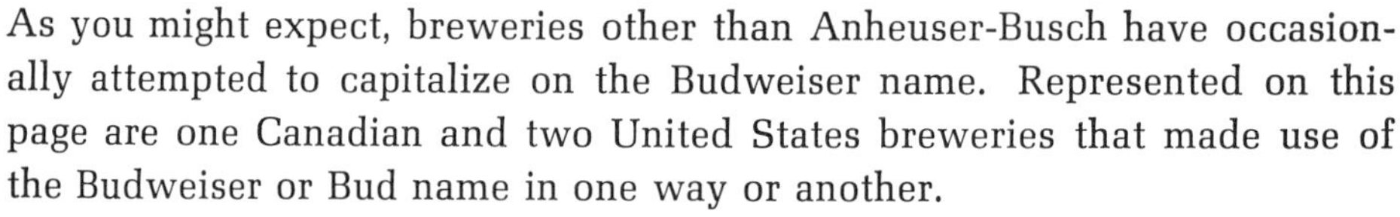

As you might expect, breweries other than Anheuser-Busch have occasionally attempted to capitalize on the Budweiser name. Represented on this page are one Canadian and two United States breweries that made use of the Budweiser or Bud name in one way or another.

Little is known about Canada Bud (upper right tray), but it is logical to assume that the "Bud" part of the brand name was not entirely original!

The second infringement was made by the Budweiser Brewing Co. of Brooklyn, N. Y. (bottom right bottle). This company operated under the Budweiser name from 1885 until 1898. In that year, rather than defend an Anheuser-Busch suit, it changed its name to the Nassau Brewing Co. (second bottle from right).

Far and away the most involved "other Budweiser" was brewed by the DuBois Brewing Co. Founded in the small western Pennsylvania town of DuBois in 1897, this company started using the Budweiser brand name in 1905. Although it had several other brands, as evidenced by the three trays and numerous cans shown here, DuBois did use Budweiser as one of its major labels until 1970. Effective October 31st of that year, it was prohibited from using the Budweiser name by a Federal court order. DuBois, owned by the Pittsburgh Brewing Co. since 1967, has since (1972) ceased brewing operations.

BARTELS BREWING COMPANY
EDWARDSVILLE, PA.
and SYRACUSE, N. Y.

In 1886 John Greenway, already famous as an ale brewer in the "Salt City," founded a second Syracuse brewery to manufacture lager beer. He called it the Germania Brewing Co., and operated it until 1893, when Herman Bartels became the new owner. Ironically, the first thing Bartels did was to add ale and porter to the brewery's product line.

Bartels, born in Richtenberg, Prussia in 1853, was involved with numerous breweries. He came to America in 1872, and worked for several different New York City breweries. In 1878 he became part owner of the Crescent Brewing Co., of Aurora, Indiana. Here he remained until 1884, when he sold his interest in Crescent and moved to Cincinnati, where he purchased an interest in the J. Walker Brewery. He moved again, this time to Syracuse, in 1887, when he purchased Germania. He worked there as brewmaster for The Haberle Brewing Co. until 1893. Still not content after buying Germania, Bartels later became part owner of the Bartels Brewing Co. of Edwardsville, Pa., and the Monroe Brewing Co. of Rochester, N. Y.

Both Bartels' plants re-opened following Repeal. The Syracuse company brewed Devonshire Ale, Crown Lager and India Pale Ale & Porter. It became entirely independent in 1940, but remained in operation only a short time after that—closing down in 1942. The Edwardsville facility, which brewed Bartels Beer, lasted until 1968, when it, too, ceased operations.

John Greenway's Syracuse Ale Brewery, which he founded in 1853, the year Herman Bartels was born. This ad is from the 1868/69 Onondaga County Directory.

PREMISES OF THE BARTELS BREWING COMPANY, SYRACUSE, NEW YORK.

PREMISES OF THE MONROE BREWING COMPANY, ROCHESTER, NEW YORK.

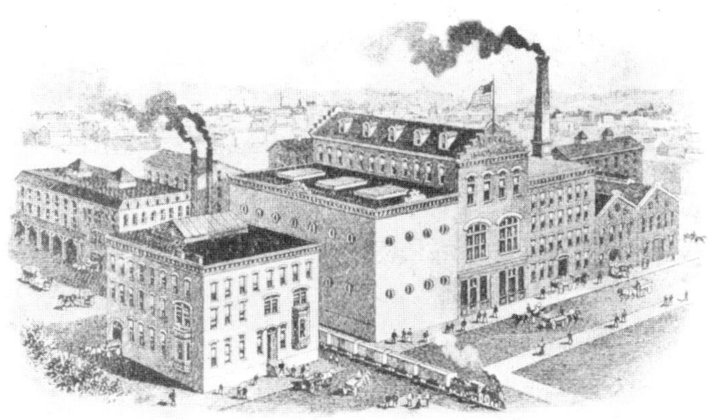

PREMISES OF THE BARTELS BREWING COMPANY, EDWARDSDALE (NEAR WILKESBARRE), PENNSYLVANIA.

JAMES R. NICHOLSON, President

A wide variety of Bartels' items. While most of the collectibles shown are from Edwardsville, the right-hand tray, as well as four of the embossed bottles, are from Bartels' "home brewery" in Syracuse. Also pictured is a blob bottle from Bartels' third brewery, the Monroe Brewing Co. of Rochester.

SYLVESTER PAUKSZTIS, Superintendent

PETER JENTZER, Brewmaster

The April, 1909, issue of the *North American Wine and Spirit Journal* featured an article entitled, "Brewery Success—The Remarkable Growth and Development of the Bartels Brewing Company." This was only the story of the Edwardsville company, and contained about 99% verbiage, and 1% meaningful fact. It did, however, contain some interesting pictures of the brewery's personnel. Note that Herman Bartels is not included here. Since Bartels died a year later, in 1910, it's logical to assume that by 1909 he had withdrawn from the Edwardsville operation because of poor health.

BEVERWYCK BREWING COMPANY

30-52 NORTH FERRY STREET
ALBANY, N. Y.

The Beverwyck Brewing Co. was an important New York State brewery for many years. It was founded in 1878 as the lager-beer offshoot of the Quinn and Nolan Ale Brewing Co. (originally founded in 1845). The two breweries were side by side on North Ferry St. Although the ale brewery was successful in its own right and outsold its younger brother for quite a few years, the popularity of lager finally won out and by 1903 it could be stated that "the capacity of the ale plant is 80,000 barrels per annum, that of the lager-beer brewery, 125,000 barrels."

After Repeal the operations of the two breweries were combined, with both lager and ale produced. In 1950 the brewery was closed and the plant sold to the F. & M. Schaefer Brewing Co. of Brooklyn, N. Y. Schaefer, a lager brewer, continued to brew Irish Cream Ale for awhile, but then dropped it, thus eliminating the last Beverwyck vestige. With the opening of their new Lehigh Valley, Pa., plant in 1972, Schaefer found themselves with excess capacity, and the older and less efficient Albany plant was therefore shut down as of December 31 of that year.

Beverwyck's spout cans are of special interest because they highlight, perhaps better than the output of any other brewery, the minor variations that are possible in can collecting. At first glance, for example, the three Irish Cream Ale spout cans on the right look to be the same. A second look will evidence quite a few differences. The can to the right states, "Irish Cream Ale *Brand*," whereas the middle can reads, "Irish *Brand* Cream Ale" and also includes the "BBBB" logotype (Beverwyck Best Beer Brewed). The can farthest left is even more interesting—it is exactly the same as the first can, except that it states, "Irish Cream Ale *Type*" instead of *Brand*. This change must have been made for some kind of legal reason.

Even more puzzling are the two cans in the middle-front. They look the same except that the can to the right is an "Export" can. Once again, however, there are differences to be seen. Beverwyck was founded in 1878 and that is the date given on five of the cans, including the left front can. However, the right front can (as well as the olive drab Army can) states "Since 1845," which was Quinn & Nolan's founding date.

BLITZ-WEINHARD COMPANY

1133 WEST BURNSIDE STREET
PORTLAND, OREGON

Blitz-Weinhard is the oldest continuously operating brewery west of the Mississippi, and has been Oregon's only brewery since 1952.

Henry Weinhard was born in Lindenbronn, Württemberg, Germany, on February 18, 1830. As a teenager he learned the brewing trade in Germany and then emigrated to America when he was 21, in 1851. He worked for numerous breweries in Philadelphia, St. Louis and Cincinnati, before heading for the greener pastures of the West Coast in 1856. Along with his German brewmaster's papers and a handmade copper brew kettle, Weinhard, after brief stops at Sacramento and San Francisco, finally stepped ashore at Fort Vancouver (now Vancouver), Washington.

Weinhard soon heard reports of the Cayuse Indian Wars occurring in eastern Oregon. With an eye on the Fort's 300 soldiers as protection and potential customers, he set up his first brewery just outside the fort's walls. Right from the first batch brewed in that small homemade kettle, business was good, and within a year wooden kegs filled with Weinhard's beer, and proudly branded "HW," were being shipped as far east as Fort Walla Walla.

In 1862, when a record $20,000,000 taken from the eastern Oregon gold mines helped turn Portland into a bustling boom town, Weinhard decided to move his brewery across the river to that city, where it's been ever since. The "Blitz" part of the company name came by way of Arnold Blitz. In 1909, Blitz purchased a Weinhard competitor, the Portland Brewery, and operated it until Prohobition. Then, between 1924 and 1928, during the very dry days of Prohibition, he optimistically rebuilt the entire plant.

In 1928 the Blitz and Weinhard breweries merged, forming today's Blitz-Weinhard Company.

As this beautiful old sign proudly proclaims, the Weinhard Brewery won three Gold Medals at the Lewis & Clark Exposition, held in Portland in 1905. Unfortunately, while planning for this great centennial, Henry Weinhard passed away in 1904.

EAGLE

The eagle, both in name and symbol, has always been closely associated with American brewing. There have been at least ten different Eagle breweries through the years and this stately bird has also adorned the advertising and packaging of many other United States breweries (most notably, of course, Yuengling's and Anheuser-Busch).

Collectibles from eight "Eagle" breweries can be found: Lembeck & Betz Eagle Brewing Co., Jersey City, N. J. (1869-Prohibition); Eagle Brewing Co., Newark, N. J. (1892-1919); Elgin Brewing Co., Elgin, Ill. (1849-1919); Eagle Brewing Co., Utica, N. Y. (1888-1943); Eagle Brewing Co., Providence, R. I. (1899-1919, but plant dates to 1873 as Bartholomew Keily Brewery); Eagle Brewing Co., Catasaqua, Pa. (1934-1963); Eagle Brewing Co., Waterbury, Conn. (c. 1903-1919); L. H. Roemer & Co. Eagle Brewery, New York, N. Y. (1880-1892).

Shown with the five cans and two labels from the Eagle Brewing Co. of Catasaqua are two embossed bottles from the pre-Prohibition Catasaqua Brewery. Catasaqua is a village near Allentown, in eastern Pennsylvania.

PREMISES OF THE LEMBECK & BETZ EAGLE BREWING COMPANY, JERSEY CITY, NEW JERSEY.

The name Betz, which appears in two different places on these two "Eagle" pages, is one that pops up constantly in any research on Philadelphia, New York City, and New Jersey breweries.

John F. Betz, who had learned the art of brewing at the Yuengling Brewery in Pottsville, Pennsylvania, opened a brewery at 347-355 W. 44th St., in New York in 1853. He operated this brewery, known as the Eagle Brewery, until 1880, when he sold it to L. H. Roemer. Betz had also, in 1867, started a brewery in Philadelphia, on New Market and Callowhill Streets. Both were fairly sizeable plants. In 1878, for example, Betz's Philadelphia brewery sold 52,891 barrels (third among the 85 breweries in operation in the Quaker City that year), while the New York plant sold 28,961.

In 1880, when disposing of the New York brewery, Betz moved his Philadelphia operation to a new location, the junction of Crown, Willow, 5th and Callowhill Streets. There he added lager beer to his former production of ale and porter. In 1886 he became involved with a second concern in Philadelphia, the Germania Brewing Co. This he operated until around the turn-of-the-century when it was sold to Henry Hess. In August, 1897, Betz and his son John, Jr., returned to New York, purchasing the former David G. Yuengling, Jr., brewery at 10th Avenue and 128th Street. This they operated as Betz's Manhattan Brewery.

The Betz of the Lembeck & Betz Eagle Brewery in Jersey City may or may not be related to John F. Betz. This Betz was also named John and it's known that he had previously worked at John F. Betz's Philadelphia brewery before founding, with Henry Lembeck, their Jersey City brewery in 1869.

The name Betz has appeared in several other brewery listings: Betz, Yuengling & Beyer, Richmond, Va.; Elias & Betz, 403 E. 54th St., New York, N. Y.; John J. Betz, 9th Ave. & 60th St., New York, N. Y. There was even a short-lived (1934-1939) post-Prohibition brewery called John F. Betz & Son, in Philadelphia at 415 Callowhill St.

ESSLINGER BREWING COMPANY

417 N. 10th STREET
PHILADELPHIA, PENNSYLVANIA

George Esslinger founded this long-lasting Philadelphia brewery in 1868. His first address was 1012 Jefferson Street, but he later moved to a larger plant on North 10th Street. Esslinger's came back strong after Prohibition, and the Esslinger "Little Man" became a familiar scene on posters and ads around Philadelphia. The company did so well that for many years (1937-1953) it operated two breweries, a lager plant at 417 North 10th St., and an ale plant at 2205 N. American St. Bad times eventually set in, however, and Esslinger's closed down in 1964.

Note the two bottles—illustrating two different periods, before George Esslinger's son was admitted to the firm, and after.

Gretz (can display, middle of bottom row) was the product of another venerable Philadelphia brewery. Founded as Rieger and Gretz in 1881, it reopened after Prohibition as the Wm. Gretz Brewing Co., and remained in operation until 1961. Esslinger bought the rights to the brand name and brewed it for the three years in which they outlived Gretz.

FALSTAFF BREWING CORPORATION

1920 SHENANDOAH AVENUE
ST. LOUIS, MISSOURI

Falstaff traces its lineage back to 1917. In that year, with Prohibition looming up as a dark shadow, Joseph "Papa Joe" Griesedieck purchased the small Forest Park Brewing Co. plant and formed his own firm, the Griesedieck Beverage Co. Soon Prohibition did take effect, but the small company managed to stay in business by producing soft drinks, near beer, and even smoked hams and bacon. During Prohibition the name of the firm was changed to The Falstaff Corporation and, later, when it was possible to see the light of Repeal, to Falstaff Brewing Corp. Falstaff, in fact, obtained Federal Permit #1 when brewing was legalized again in 1933. Two years later, the company started (along with Pabst) the multiple-plant concept, now so prevalent in brewing, by purchasing the former Fred Krug Brewery in Omaha, Nebraska. Prior to Falstaff's success in Omaha (and, a year later, New Orleans) it had been thought impossible to brew identical beer in different and widely separated plants.

Falstaff is now among the top ten American breweries in sales, and operates eight plants located throughout the country. Included is the Narragansett Brewing Co., of Cranston, Rhode Island, a very sizeable New England brewery purchased in 1965. Another proud beer name, Ballantine, is also now in the Falstaff stable, as Falstaff acquired the rights to the brand when P. Ballantine & Sons went out of business in 1972.

Of perhaps more interest to collectors is the fact that Falstaff operates a Museum of Brewing in St. Louis, although this museum is now open by appointment only. Several fine trays from the museum are pictured in the color pages as well as on this page.

BRANCH AT SAN FRANCISCO, CAL.

The relationship between Lemp, Griesedieck and Falstaff has always been confusing. Griesedieck Bros. was formed in the early 1900's by Joseph Griesedieck's brother, Henry, and Henry's sons, Anton, Henry and Raymond. It merged with Falstaff in 1957. It was from Lemp that Joseph Griesedieck purchased the Falstaff trademark and shield in 1920. Notice that the outline of the embossed Lemp label is exactly the same as Falstaff's can design. That's the Falstaff "shield." The name "Falstaff," of course, refers to Sir John Falstaff, Shakespeare's immortal comic character. Through the years this jolly 15th-century knight has become synonymous with the thought, "eat, drink and be merry," and was, as mentioned, originally the trademark of the William J. Lemp Brewing Co., direct descendant from one of St. Louis' pioneer brewers, J. Adam Lemp. Lemp established his original brewery in 1840, and operated it for 22 years until his death in 1862, when the business was inherited by his son, William J. Lemp. Under the son, Lemp became one of the first "shipping" or national breweries. Branches, agents and depots were important to a brewer striving to "go national." By 1893, when they issued a Souvenir Booklet for visitors to their exhibit at the Chicago Columbian Exposition, Lemp had a remarkable 230 depots spread over 29 states. Shown here are some of those depots.

BRANCH AT HOT SPRINGS, ARK.

BRANCH AT LEADVILLE, COLO.

BRANCH AT LOS ANGELES, CAL.

BRANCH AT BIRMINGHAM, ALA.

CHRISTIAN FEIGANSPAN

50 FREEMAN STREET
NEWARK, NEW JERSEY

This is another world-famous brewery name. What reader has not heard of P.O.N.—Pride of Newark, Feiganspan's brand name for over 50 years? Always an extremely large brewery by itself, Feiganspan also gained controlling interests in the Dobler Brewing Co. of Albany, N. Y., and the Yale Brewing Co. of New Haven, Conn., shortly before Prohibition. By 1939 the brewery could boast that P.O.N. was the largest selling ale in New York City (they also produced P.O.N. Lager, Porter and Stout) and that the plant could bottle and/or can over 1,700,000 units a day. The wars years were not kind to the company, however, and by 1944 Feiganspan had been bought out by its down-the-block competitor, P. Ballantine & Sons.

Dobler, located on South Swan Street in Albany, actually outlasted its parent company by sixteen years, staying in business until 1959. The Dobler trade name is still in use today, by Piel Brothers, in their Willimansett, Mass, brewery.

The Yale Brewing Co. lasted a considerably shorter period of time. Founded in 1902 (it had earlier, since 1885, been known as the Quinnipiac Brewing Co.), it closed because of Prohibition, and never reopened.

The tray showing a girl with flowing hair was issued by both Feiganspan and Dobler, the only difference being the name of the brewery printed on the inside top rim.

Ballantine re-activated the Feiganspan name in the 1960's as the brewery of record for its budget-priced Munich beer, but Ballantine, too, went out of business in 1972. Its brand names were bought by Falstaff, who has continued to market Munich, still using the Feiganspan name on the label.

FIDELIO BREWING COMPANY

1st AVE., 29th to
30th STREETS
NEW YORK, NEW YORK

Founded in 1852, this brewery was known as H. Koehler & Co. until 1917, and its brand of beer was "Fidelio." Shortly before Prohibition, the company name was changed. During Prohibition, Fidelio manufactured cereal beverages and then, following Repeal, brewed Fidelio, McSorley and New Yorker. The brewery switched names in 1939, becoming the Greater New York Brewing Co., but continued the use of the same brand names. Operations ceased in 1943.

While the drinking scene and McSorley's trays appear to have been issued prior to Prohibition, they were not. Both have 1935 copyright dates on them.

The three embossed Koehler bottles naturally pre-date 1917; the labeled bottle is from the post-Prohibition Fidelio brewery.

GENESEE BREWING COMPANY

(including Bartholomay Brewing Company)

100 NATIONAL STREET
ROCHESTER, NEW YORK

This well-known Rochester, N. Y., brewery dates back to 1878. From that date until 1889 it operated independently as the Genesee Brewing Co., selling its Liebotschaner Beer throughout western New York State. In 1889, however, it became part of the Bartholomay Brewing Co. complex (consisting of the Bartholomay Brewery, Rochester Brewing Co., Genesee Brewing Co., E. B. Parsons Malting Co., and the S. N. Oothout & Son Malting Co., all of Rochester).

In 1933 Genesee reopened as a re-organized and again independent company. The brewery's goal that year was to sell 75,000 barrels; by 1972 annual sales were over 1,700,000 barrels. Genesee now markets its beer in eight states, and ranks in the top twenty largest U. S. breweries.

Both trays shown are from the 1930's. The 12-horse team shown on the right-hand tray, as well as on the 12-Horse Ale labels and the framed poster, was part of Genesee's marketing program from 1933 until 1950.

All of the cans carry obsolete designs. Spouts were discontinued by Genesee during W.W. II, while Dickens Ale was only brewed from 1957-59.

Bartholomay, of Vincent Place, Rochester, once the parent company of Genesee, was founded in 1852 by Philip Will and Henry Bartholomay. After five years, Will left the partnership and Bartholomay carried on alone. In 1874 the company was incorporated and it was incorporated again in 1889 when, as mentioned, it became the controlling name for three Rochester breweries and two malting houses.

Knickerbocker, a brand name generally associated with New York City's Jacob Ruppert, was actually first used by Bartholomay in 1892. While the two other breweries it absorbed in 1889 (Genesee and Rochester) both reopened after Prohibition, Bartholomay did not.

B. B. CO. DINING ROOMS.—COTTAGE HOTEL.

BARTHOLOMAY BREWING CO. COTTAGE HOTEL,

It was a pre-Prohibition tradition of many breweries to own and operate saloons, restaurants and hotels—where naturally only their own products were sold. While this practice was made illegal under the terms of Repeal, many of these "tied houses" were quite grandiose during the time the practice was allowed. Such was the case with Bartholomay's Cottage Hotel and Pavilion, as described and pictured here in a booklet put out by the brewery in the late 1880's.

Some early Bartholomay bottles. As can be seen on the two bottles to the right, the company's symbol was a rather unusual one—a wheel with two outstretched wings.

An ad placed by Bartholomay's Boston distributor and bottler.

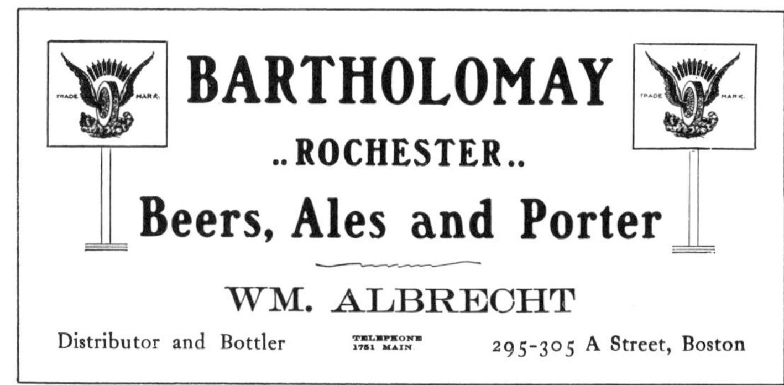

GETTELMAN BREWING COMPANY

14th and PRAIRIE STREETS
MILWAUKEE, WISCONSIN

The building of this brewery was originally undertaken in 1854 by two gentlemen named Strohn and Reitzenstein. Work had barely passed the foundation stage when both were overtaken by a cholera epidemic and died. The advantages of the site, though, were not long overlooked and George Schweickhart, who had just moved to Milwaukee from Buffalo, purchased the plot and completed the plant. It was thus operated until 1876 as the Schweickhart (or Menominee) Brewery. In that year the company was bought by Adam Gettelman, Schweickhart's son-in-law. The name, however, was not changed to the A. Gettelman Brewing Co. until 1887.

The famous $1,000 Beer originated back in 1891 when the company started bottling beer and a trade name was needed. Gettelman thought up the $1,000 Beer name and offered this amount to anyone who could prove that anything other than pure malt and hops had gone into his product. In later years this brand name was heavily promoted through the slogan, "Gettelman—Wisconsin's One Grand Beer!"

Gettelman survived Prohibition by brewing a near bear. A smaller brewery that basically distributed only in Wisconsin and Northern Illinois, it was bought by the Miller Brewing Co. in 1961. Miller operated Gettelman as a wholly-owned subsidiary until 1972 when the plant was closed. The use of the Gettelman brand name, however, has been continued.

Various Gettelman collectibles. If you look over at the lower right Milwaukee's Best can closely you will see a 2¢ beer tax stamp to the left in the "B" in "Beer." The stamp was affixed by the city of Brunswick, Ga., where this can was purchased in the early 1960's—the only city beer tax stamp I've ever seen.

HABERLE CONGRESS BREWING COMPANY

BUTTERNUT &
McBRIDE STREETS
SYRACUSE, NEW YORK

Of the numerous breweries "The Salt City" has had over the years, this was the most famous and longest-lasting. Founded in 1857 by Benedict Haberle, it was for many years a very small brewery. In 1879, for example, Haberle sold but 4607 barrels, which placed it 139th on the list of New York state's 339 breweries.

Beginning in 1892, however, Haberle began to grow dramatically by acquisition and consolidation. First, in 1892, came a merger with the Crystal Spring Brewing Co. (founded in 1887 and located on Burnet Ave. and Vine St.), to form the Haberle Crystal Spring Brewing Co. Then, in 1900, two other Syracuse breweries, Wm. Kearney's Sons and the National Brewing Co., were bought out.

After Repeal the company operated as the Haberle Congress Brewing Co., with only the original Haberle plant on Butternut Street in operation. Operations ceased December 31, 1962.

The George Ringler calendar is a most colorful example of American breweriana. Ringler's Brewery was founded in 1872, and was one of the many victims of Prohibition, closing in the 1920's after vainly trying to market near beer.

Situated adjacent to Ruppert's, this once beautiful plant has been torn down and the land on which it was situated is now included, ironically, in the Jacob Ruppert Urban Renewal project.

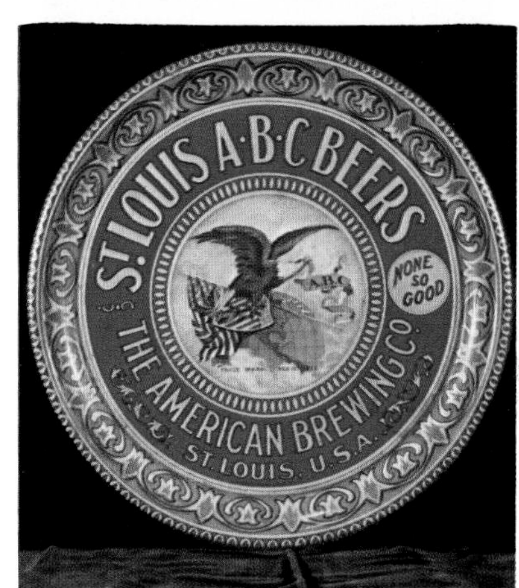

The three beautiful St. Louis trays are examples of the great material on display at Falstaff's Museum of Brewing, St. Louis, Mo.

To show these four Lion items in anything but color would be a crime.

Lion Tonic was a product of Obermeyer & Liebmann, a Brooklyn brewery founded in 1868 and merged into S. Liebmann's Sons (now Rheingold) in 1924.

Because of the very obvious canal scene shown on both the 1889 and the 1895 calendars, Windisch-Muhlauser looks as if it were located in Venice rather than in Cincinnati. Known as the Miami and Erie Canal, the waterway was visible until 1928 when it was covered over by construction of the Central Parkway. The Burger Brewing Co. purchased this beautiful old (built in 1866) brewery in 1943, and utilized it until they went out of business in 1973.

The calendar of New York City's Lion Brewery is from 1914, a good year in which to identify with the Panama Canal. It was at this time, after eleven grueling years, that the United States finally finished one of mankind's wonders of the world.

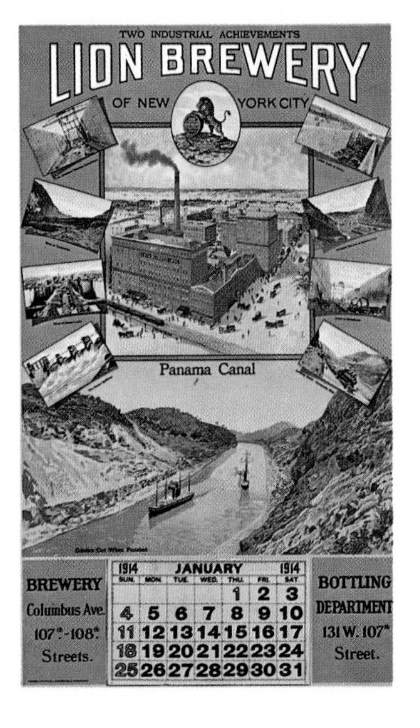

Illustrated here are some of the most colorful brewery trays from representative firms, as well as a 1911 calendar from the Consumers Brewing Co. of Brooklyn, N.Y. In spite of the name, Consumers was actually located on Betts Ave., in Woodside, Queens—not Brooklyn.

It's interesting to note the way in which the plant majestically towers over all of Queens and Brooklyn—a veritable giant among breweries!

The award for the individual brewery's most colorful collectibles shown in THE BEER BOOK must go to D. G. Yuengling & Son, Pottsville, Pa. Fittingly, Yuengling is also the oldest operating brewery in the United States.

The eagle has always been a symbol of strength and character—and therefore adorned many a beer tray. This is a 4 1/8'' change tray.

The poster shows the brewery as it looked circa 1880.

Yuengling, to this day, brews porter—one of the few United States breweries that still does.

The most colorful of representative breweries' cans. As can be seen here (and as can collectors well know), sets from the same brewery can be strikingly beautiful with respect to color and contrast.

The Eberhardt & Ober poster is surely one of the best ever produced. It is colorful and detailed beyond imagination. Notice the streetcar, the beautiful park and fountain across from the brewery, and the three flags proudly flying. This, of course, is an artist's rendition and, therefore, somewhat suspect with respect to accuracy. There was no attempt, however, to disguise the black pollution pouring from the plant's smokestack.

Eberhardt & Ober was one of 21 Pittsburgh area breweries that were consolidated in 1899 to form the Pittsburgh Brewing Co. The red sign directly under the poster is from this merged firm and already the E&O had come to stand for "Early & Often" as much as for Eberhardt & Ober.

Breweries received inspiration for their brand names from many sources, but the use of "Nickel Plate" by Centilivre of Fort Wayne, Ind., is the only instance that comes to mind when a beer was named after a railroad. Nickel Plate is the nickname of the New York, Chicago & St. Louis Railway, a line which served Fort Wayne.

Globe was an important brewing name in Baltimore for many, many years. Though the company's beginnings go back to 1816, it was the Wehr-Hobelmann-Gottlieb Brewing & Malting Co. that first adapted the name Globe Brewery, in 1888. The beautiful Globe Bottled Beer sign shown had to be produced between that date and 1899; in March of 1899 Globe and 16 other Baltimore breweries combined to form the Maryland Brewing Co., Inc. (reorganized two years later as the Gottlieb-Bauernschmidt-Straus Brewing Co.).

Nashville has become synonymous with country and western music, but in 1897 it meant the Exposition and Hauck's Beer.

Select was the predecessor of Pabst Blue Ribbon, Milwaukee.

Anheuser-Busch's St. Louis ("Covers 70 City Blocks") as portrayed in a 1914 booklet, "Epoch Marking Events of American History."

Two typical turn-of-the-century brewery tour souvenir booklets.

THEODORE HAMM COMPANY
720 PAYNE AVENUE
ST. PAUL, MINNESOTA

While Hamm is today one of the top ten largest brewing companies in America, its beginnings were small indeed. The original plant, on the heights overlooking St. Paul, was built in 1864 by A. F. Keller. A year later, in 1865, Theodore Hamm purchased the brewery from Keller. Hamm was a native of Baden, Germany, where he was born in 1824. At the age of 19, in 1854, he emigrated to America. His original trade was that of butcher, and he worked as a butcher in Buffalo and Chicago before settling down in St. Paul in 1856. In St. Paul he ran a boarding house and saloon until his purchase of the brewery from Keller.

In its early years the capacity of the plant was but 500 barrels a year—contrasted with the 14,000 barrels that Hamm can now produce in a single day.

In 1953 Hamm expanded to the West Coast by buying the plant of the Rainer Brewing Co. in San Francisco. Four years later, it also added a plant in Los Angeles (since closed, in the spring of 1973).

THEODORE HAMM, ST. PAUL, MINNESOTA.

The blob-top quart and 12 oz. bottles were the type used from around 1880 to 1903, when Hamm installed its first capping machine and made the switch to crown-top bottles.

As with most all pre-Prohibition breweries, Hamm had its own horses and wagons to deliver its kegs of sparkling lager. The round metal object in front of the bear is part of a horse's livery.

Of the cans, the two oldest are the bottom row, left (used up to 1951) and the olive-drab military can (bottom row, right, probably used from 1940 to about 1949). The can in the bottom row, middle, is of the 1970's but has an appealing, old-time look to it because of the brewery scene and typeface used. The can on the top row, right, was used from 1951 to 1957. The last remaining can, top row, left, is one for malt liquor and the Heublein name reflects the fact that since 1969 Hamm has been owned by Heublein, Inc.

As part of their "Born in the Land of Sky Blue Waters" advertising theme, the Hamm's bear has become a well-known symbol. Hamm produced a limited number of Hamm's Bear Collector's Decanters (filled with beer, incidentally) for the 1972, Christmas season.

THE JAMES HANLEY COMPANY

35 JACKSON STREET
PROVIDENCE, RHODE ISLAND

Hanley's beginnings are somewhat involved, going way back to 1835. In that year Otis Holmes built a small brewery at the corner of Jackson and Fountain Streets in Providence. Here he brewed until 1862, when (probably because of death) the plant was closed. It remained closed for five years, until 1867, when a John Bligh bought the property and plant. He enlarged and expanded the plant, and operated it until his death in 1875. Enter James Hanley and John P. Cooney who ran the company as a partnership for three years under the name of Cooney & Hanley. In February of 1879 Hanley bought out his partner and changed the name to James Hanley & Co. In 1885 Hanley, for some unknown reason, changed the name of his company to the Rhode Island Brewing Co. Finally, in 1896, it evolved back to the James Hanley Brewing Co.

Thus it remained—except for the Prohibition years—until 1958, when the company went out of business. The Hanley brand name, however, still lives on as Narragansett obtained the rights to the name and has continued to use it quite heavily.

It appears that "The Connoisseur" was Hanley's symbol in the pre-Prohibition years. He's shown twice here—on a standard tin tray and on an enamel tray. After Prohibition a complete switch was made and the Hanley bulldog became widely known as Hanley's "Quality Guarded" symbol.

The bold statement on the bulldog tray, "Brewed in Rhode Island's Largest Ale Brewery," calls for comment. Since James Hanley was an Irishman (he came to the United States at the age of four), he undoubtedly took great pride in his ale—but still, how many ale breweries could Rhode Island have had? Actually, this was a subtle way of saying "We sell more ale than our competitor, Narragansett!"

HARVARD BREWING COMPANY

24 PAYTON STREET
LOWELL, MASSACHUSETTS

This was truly a great among New England breweries—and it only seems proper and fitting to do it justice by quoting directly from *One Hundred Years of Brewing.*

"This company was incorporated in 1898, the plant being located on the New York, New Haven & Hartford and Boston & Maine railroads in the southern outskirts of the city of Lowell, and consists of two separate and complete breweries, one for lager beer and the other adapted to the brewing of ale and porter, a mammoth bottling house, magnificent office building, stables, cooper shops, carriage houses, etc. The annual capacity of the plant is three hundred and fifty thousand barrels.

The products of the company are sold throughout the Eastern, Central and Atlantic Coast States. The plant is the largest in the New England States, and is considered to be one of the finest and best equipped breweries in the United States."

Harvard reopened after Repeal, operating independently until 1957. In that year the company merged with the Hampden Brewing Co. of Willimansett, Mass., becoming Hampden-Harvard Breweries, Inc. The Lowell plant was closed and all operations conducted out of Willimanset. The Harvard trade name, however, was carried on and, in fact, still is, even though Hampden-Harvard became part of the Piels (Associated) combine in 1962.

A classic (and very picturesque—see color pages) item, indeed, is the "Charles Gibson Dana Girl" tray pictured at upper left. The tray in the upper right is a 1907 calendar Vienna Art Plate. The months of the year are printed around the outside rim. The large tray is post-Prohibition and of aluminum.

It was almost a New England tradition to date bottles—and Harvard was not about to break tradition. What's puzzling is that the '98 bottle is a crown top, while several of the more recent years are blob. It should be the reverse. The back, right, 1902 bottle should be noted for its unusual shape; it appears to be shaped more like a wine than a beer bottle.

The Vienna Art Plate is courtesy of America Hurrah Antiques, New York, N. Y.

Ad from April, 1909, *North American Wine & Spirits Journal.* Note the impressive list of agencies.

CHRISTIAN HEURICH BREWING COMPANY

25th, 26th, D and
WATER STREETS, N.W.
WASHINGTON, D. C.

The nation's capital has always been much more of a center for government than for commerce, and never much of a brewing center. In 1898, for example, there were only six breweries operating, in contrast to 26 in close-by Baltimore.

The one Washington brewery that did gain considerable sales and fame was Heurich. Its beginnings go back to 1864, when George Schnell started a small Weiss beer plant on 20th Street, N.W. In 1872 Christian Heurich (and Paul Ritter, who was a partner for a very short time) bought the brewery and virtually rebuilt it. In spite of enlarging and improving it, however, the plant remained a moderate-sized facility with a capacity per year of but 30,000 barrels. Heurich was not satisfied and in 1895 he constructed an ultra-modern, 500,000-barrel-capacity plant at 25th and Water Streets. Built entirely of brick, stone and concrete, Heurich carried no insurance on the building—undoubtedly the only large brewer in the country that could make that statement.

After Prohibition, only two Washington breweries reopened, Heurich and Abner-Drury. Abner-Drury, located at 2424 G Street, N.W., became the Washington Brewery in 1937, and went out of business a year later. This left Heurich as the District of Columbia's last brewery, an honor it upheld for 16 long years, until it, too, closed in 1956.

CHRISTIAN HEURICH, WASHINGTON, D. C.

A pre-Prohibition bottle from Heurich's Norfolk, Va., branch, plus a 1934 bock label and a small Heurich can display.

BREWERY PREMISES OF CHRISTIAN HEURICH, WASHINGTON, DISTRICT OF COLUMBIA, 1902.

A good sampling of some of the many types of beautiful signs put out by a brewery. The two large items are definitely pre-Prohibition and were for saloon use. The diamond-shaped sign is also pre-Prohibition and unusual because of its shape. Maerzen was probably a specially brewed, dark beer. It is also mentioned on the black large saloon sign. The small tin, "Thank You–Call Again" sign is from the post-Prohibition period. It reflects the increased importance breweries began to place on in-home consumption. Bottle (and can) sales grew by leaps and bounds, and *store* point-of-purchase became as important as *saloon* point-of-purchase signs formerly had been.

JOHN HOHENADEL BREWERY

CONRAD STREET and INDIAN QUEEN LANE PHILADELPHIA, PENN.

Back in 1880, when John Hohenadel established this brewery in a picturesque setting overlooking the Falls of the Schuylkill River, the output for the year was but 800 barrels. By the early 1900's that figure had increased to 20,000 barrels, but Hohenadel was never a large brewery by any standards. It did, however, stay in business through most of the Prohibition years, producing cereal beverages.

After Repeal, brands included Hohenadel Beer, Alt Pilsener, Health Porter and Indian Queen Ale. Hohenadel went out of business in 1953.

The rectangular tray is very much pre-Prohibition, while the three round trays are from the 20-year period during which Hohenadel operated after Repeal. Note that the name was hyphenated before Prohibition (on both the rectangular tray and the mug). This must have been to help people pronounce it correctly; you couldn't order their brand of beer if you couldn't pronounce the name.

JACOB HORNUNG BREWING COMPANY

3111 NORTH 22nd STREET
PHILADELPHIA, PENN.

Jacob Hornung founded this company, originally called the Tioga Brewery, in December of 1885. At first he brewed only lager, but later he also added porter, ale and his famous White Bock to the line. Never a very large brewery, Hornung did open after Prohibition and stayed in business until 1954. The White Bock tray is probably pre-Prohibition, issued in 1912 or 1913 to capitalize on the "Grand Prix" Hornung's brew was awarded at the Paris International Exhibition of 1912.

The round "Beer That Wins Awards" tray has a copyright date of 1946, and features not only the Paris honor, but also a First Prize awarded by the New Jersey Licensed Beverage Association in 1934. The New Jersey award is further explained on the back of each of the Hornung cans: "60 expert beer tasters appointed by the Association in a hidden identity test, awarded Hornung's first prize over 24 national and local brands by a vote of more than 2 to 1." I wonder how one qualifies as a "beer taster"?

CHARLES D. KAIER COMPANY

67-79 N. MAIN STREET
MAHANOY CITY,
PENNSYLVANIA

In 1862, Charles D. Kaier, a native of Biningan Baden, Germany, founded a small brewery in Mahanoy City, in Eastern Pennsylvania. Through the years it prospered and grew, becoming a perfect example of how a sizeable (100,000 barrels-per-year-capacity by 1912) brewery could enrich a community. Kaier, himself, was extremely community-minded, and was involved in all sorts of local civic-oriented activities. He was also very patriotic. When the Spanish-American War broke out, he promised his entire working force that he would continue to pay any or all of them full wages as long as they were in their country's service in the field of action!

Probably because it was apart from any major population center, Kaier was much more self-sufficient than most other breweries of the day. The company had its own cooperage shop, blacksmith shop, carpenter shop, tinsmith shop, plumbing shop, the largest individual ice plant in the area, and manufactured its own wagons and harnesses.

Charles D. Kaier died in 1899, at the age of 60. He was succeeded as President of the company by his wife of 36 years, Margaret. Kaier re-emerged after Prohibition, operating as an independent brewery until 1966. In that year it was sold to the Henry F. Ortlieb Brewing Co. of Philadelphia. Ortlieb continued the plant in operation for about two more years, finally closing it in 1968. Ortlieb has, however, kept the Kaier's brand name in use, mostly through its Fuhrmann & Schmidt subsidiary in Shamokin, Pa.

MR CHAS. D. KAIER

Some post-Prohibition Kaier trays and cans, and pre-Prohibition bottles. The middle front squat blob is an early Weiss beer bottle; while the crown quart was most likely a Prohibition soda bottle.

A brewery the size of Kaier could have a very real social and cultural (as well as commercial) impact on a smaller community. These pictures are all from a booklet entitled a "Souvenir of Mahanoy City, Pa." prepared by the Charles Kaier Company for its golden anniversary in 1912.

Kaier was especially proud that he had built, for the people of Mahanoy City, "One of the safest, prettiest and best-equipped opera houses in the interior of Pennsylvania."

The old and the "new" with respect to delivery vehicles, one of which is posed in front of the local talking picture theatre featuring "The Girl From Nowhere." Prices started at 5¢, and the best seat in the house, according to the sign, was 20¢.

LION

As with the eagle, the lion has been a favorite name and symbol for brewers. There have been at least ten "Lion" breweries (including Stroh in its earlier days).

Represented here are six Lion breweries:

Lion Brewery of New York City, New York, N. Y. (c. 1900 to 1941): Two trays that picture a lion; two cans in right-front, see also 1914 Panama Canal calendar on color pages.

Lion Brewery, New Haven (actually Orange), Conn. (1901-c. 1908): Bottle on right. This brewery was known earlier as the Weidemann Brewing Co.

Woodworth & Roosa Lion Brewery: Bottle on far right. A real "mystery brewery"; where this company was located and its years of operation are unknown.

The Lion, Inc., Gibbons Brewery, Wilkes-Barre, Pa. (1890's—still in operation): Gibbons tray and four Gibbons cans. While Gibbons is the brand name used by this still-very-much-in-operation Wilkes-Barre brewery, its official name is The Lion, Inc.

Windisch-Muhlhauser Lion Brewery, Cincinnati, Ohio (1866-1919): See 1889 and 1895 calendars in color pages. This company once ranked right up there among the largest American breweries. In 1871-72, for example, it was the fifth largest brewery in the entire country.

122 NEW YORK CITY RECORD.

WENDEL'S ELM PARK

92d to 93d Sts., 9th to 10th Aves., NEW YORK.

WENDEL'S LION PARK,

108th St., 9th Ave., New York.

BERNHEIMER & SCHMID'S

CELEBRATED

LION BREWERY LAGER BEER,

USED EXCLUSIVELY.

Bernheimer & Schmid Lion Brewery, New York, N. Y. (1850-c. 1900): Old ad from 1885-86 New York Record. Ironically this was also among the nation's largest breweries. In 1871-72 it was number six nationally in production, immediately behind Windisch-Muhlhauser. Right around the turn-of-the-century Bernheimer & Schmid was dropped and the company became known simply as The Lion Brewery of New York City—the first brewery with which this write-up started.

OLYMPIA BREWING COMPANY

P. O. BOX 947
OLYMPIA, WASHINGTON

A tray issued by the Capital Brewing Co. in 1898, just two years after the company was founded.

Issued seven years later, in 1905, this tray reflects the same basic look as its predecessor, but reflects the change in company name.

Two beautiful steins produced for Olympia in 1904.

Leopold F. Schmidt, the father of the Olympia Brewing Co., was born in Germany, and came to America in 1866. He worked as a carpenter in several cities, and eventually ended up in Montana. There, in the early 1870's, he was introduced to brewing, more or less by accident. A friend of Schmidt's owned a small brewery in Deer Lodge, and asked him to manage it while he (the friend) was away on a trip. Schmidt agreed, and was so taken with brewing that he set up his own brewery in Butte shortly thereafter. Schmidt and his partner, Daniel Gamer, appropriately called their new company, founded in 1876, the Centennial Brewery. The firm was a success in Montana, and so was Schmidt. He was elected to the state legislature and, later, the Capital Commission.

While serving in this latter capacity, he and other Montana officials journeyed in 1895 to Sacramento, Carson City, Salem and Olympia to inspect these cities' state capital buildings. While visiting the Washington capital site Schmidt was greatly impressed with a plot of land in Tumwater, just outside of Olympia. Here, on the banks of the Deschutes River at the terminus of the fabled Oregon Trail, was an unbelievably pure artesian spring. It was love at first sight—and almost before you could say "It's the Water," Schmidt had sold his interest in the Centennial Brewery, moved his family to Washington, and was constructing a brewery on his newly purchased (for $4,500) land in Tumwater. Known originally as Capital Brewing Co., it commenced brewing in July of 1896, and the first bottles of Olympia Pale Export were placed on the market October 1, 1896. Bottled beer played an important role in Olympia's growth from the start. In fact, Schmidt was the first brewer on the West Coast to use the crown-top method of bottle sealing.

In 1902 the company name was changed to Olympia Brewing Co. in order to be more closely identified with the product the brewery was selling on an ever-larger scale throughout the West. As a result of this growth in sales Schmidt became one of the pioneers of the multiple brewery concept. In 1903 he established a brewery at Watcom, Bellingham, Washington and, in the next six years he bought the Salem Brewery of Salem, Oregon, and the Port Townsend Brewery in Port Townsend, Washington, and built the Acme Brewery in San Francisco. Unfortunately, however, try as they might, the brewmasters at these four locations could not produce a beer comparable to that brewed at the home plant. As a result, none of the lager produced at these four branch breweries was ever marketed as Olympia Beer.

Prohibition came early in Washington—1916. Olympia turned to the production of an apple drink named Applju, as well as jams and jellies. However, these products could stave off the inevitable for only five years, and in 1921 Olympia closed its doors. Reopening just the Tumwater facility after Repeal, the Schmidt family has built Olympia into the largest single brewery on the West Coast. 1972 production totaled 3,329,909 barrels, placing the company twelfth on the list of the largest breweries in the United States.

PABST BREWING COMPANY

917 W. JUNEAU AVENUE
MILWAUKEE, WISCONSIN

Now the third largest brewing company in the world, Pabst traces its beginnings back over 125 years to 1844. In that year, four years before Wisconsin became a state, a German immigrant named Jacob Best, and his four sons, Jacob, Jr., Charles, Phillip and Lorenz, started a brewery. Not surprisingly, they called it the Best Brewing Co. Helped considerably by the booming Milwaukee population growth, the company prospered and grew. Jacob Best, Sr., retired in 1853, and sold the business to Phillip and Jacob, Jr. Six years later Phillip became the sole owner when he bought out his brother's half-interest.

In 1862, Phillip's daughter Marie married a 26-year-old Lake Michigan steamboat captain by the name of Frederick Pabst. It is not known whether it was ill-fortune on the lakes or the urging of his father-in-law that caused "The Captain" to give up his nautical career, but, in any event, in 1864 Pabst became Phillip Best's partner. Within two years Best retired, selling the brewery to Frederick Pabst and another son-in-law, Emil Schandein.

In 1873, even though Best had long since retired, the business was incorporated as the Phillip Best Brewing Co. It was not until 1889, a year after Schandein passed away, that the company name became the Pabst Brewing Co. By this time it had passed the 500,000-barrels-sold-per-year mark and was the largest brewery in the country (fighting it out with Anheuser-Busch on a year-to-year basis). Frederick Pabst died in January of 1904 and management of the brewery passed to his sons, Fred, Jr., and Gustav. Surviving Prohibition nicely, Pabst has been among the industry's sales leaders ever since. As with the other national giants, Pabst now operates multiple plants (a concept they helped pioneer in 1934). In fact, their most recent branch plant is, by decree of the state, in Pabst, Georgia. No other U. S. brewery has had a town named after it.

Fred Pabst.

From 1869 until 1886, Best operated two plants in Milwaukee. This was the secondary facility, known as the South Side Brewery.

This great lithograph was issued in 1881 to honor receipt of medals by the company at the Philadelphia, Paris and Atlanta expositions.

The last year the Phillip Best Brewing Co. wished everyone a Happy New Year. Three months later the company name was changed to Pabst Brewing Co.

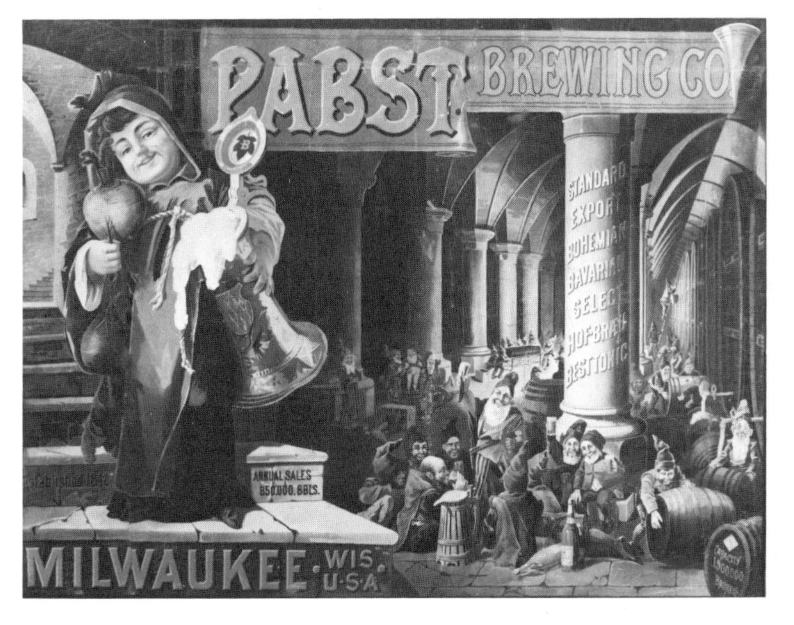

Some fabulous early Pabst advertising. All are from the early 1890's with the exception of the lithograph picturing two bottles of Blue Ribbon next to a plate of oysters. This display was originally used in November of 1903 as part of a national magazine advertising campaign to promote the Blue Ribbon brand. Magazines included *Collier's*, *Judge*, *Leslie's Weekly*, *Puck*, *Munsey's*, *Smart Set* and *Black Cat*. Eventually, the aesthetic appeal of this ad was to make it almost as successful an advertising device as the famous Victor Talking Machine Company's "His Master's Voice."

While today it's hard to say Pabst without saying "Blue Ribbon," this specific trade name was only one of several brands brewed by Pabst around the turn-of-the-century, and did not even exist until 1895.

This was Pabst's line-up in 1891. At that time "Export" was described as "Perhaps the most widely known." "Export," "Select" and "Bavarian" were sold only in bottles; "Bohemian" and "Hofbräu" in bottles or wood (kegs). There was also a brand named "Standard" that was sold only in wood.

While there is no "Blue Ribbon" trade name here, you'll notice there is a ribbon (and *it's* blue, see color pages) tied around the neck of the "Select" brand. This practice had been started in 1882, when bottled beer sales were of minor consequence to Pabst. Originally an experiment, the idea proved so successful that within ten years Pabst was buying over 300,000 yards of silk ribbon a year to tie *by hand* around the necks of the clear bottles used for "Select." The actual words "Blue Ribbon" were added to the label in May of 1895; and totally replaced "Select" in April of 1897. Three years later, in 1900, Blue Ribbon was registered and became the trademark for Pabst lager beer.

From 1900 on, Pabst has concentrated its efforts on "Blue Ribbon," as can be seen in these shots of different packaging items the company has used. The Eastside can is from Pabst's Los Angeles plant. Formerly the Los Angeles Brewing Co., it was purchased by Pabst in 1953. Notice the circled "B" trademark shown on virtually all the items. Though it's been known as Pabst for over 80 years, the company still maintains the "B," for Best, on its labels.

PEARL BREWING COMPANY

312 PEARL PARKWAY
SAN ANTONIO, TEXAS

In 1881 J. B. Behloradsky erected one of Texas' first breweries, in San Antonio. He called it The City Brewery and operated it for about two years before closing down. In 1885 a group of leading San Antonio citizens purchased the small plant, and thereby founded the San Antonio Brewing Association. One year later, the Association's manager, Otto Koehler, purchased the formula and the name Pearl Beer from the Kaiser-Beck Brewery in Germany. Actually, the German name was "Perle," to describe the little pearl-like bubbles that rise in a just-poured glass of beer. Koehler proved a

The brewery as it looked in 1886, a year after its original purchase.

Horses played a vital role in any early brewery's delivery plans. Here we see the Association's new stable, constructed in 1894.

Being "a true home industry" had to help Pearl sales. The fact that San Antonio had a very large German population helped considerably also.

good brewer-businessman and by 1916 the San Antonio Brewing Association was the largest brewery in Texas.

Then came Prohibition. During the dry years, under the leadership of Mrs. Emma Koehler, Otto Koehler's widow, San Antonio tried just about everything to stay alive—the manufacture of commercial ice, the creamery business, outdoor advertising signs, and soft drink bottling. But stay alive it did, and since Repeal this venerable San Antonio company has expanded to become the largest brewery in the Southwest.

In 1952 the company's name was changed to Pearl Brewing Company to associate it more closely with the product being sold. Nine years later, in 1961, Pearl acquired another famous old brewery—the M. K. Goetz Brewing Company of St. Joseph, Missouri. Both the San Antonio and St. Joseph breweries continue in operation today, brewing Pearl, Country Club Malt Liquor, Goetz Near Beer and Texas Pride.

A very early ad. You'll notice that "XXX" is played up and Pearl only shows on the label (where it's one word, Pearlbeer.")

The history of the "X" system of rating beer is interesting. In the 16th century when European royalty strayed far from home, a courier traveled ahead as an advance party to sample the beer in the various taverns and inns along the royal route. If the beer was but average, the courier would mark a single "X" on the inn's door; good beer earned two "X's"; while excellent beer received the highest accolade, "XXX."

Some other early Pearl advertising items. The gentlemen asking "Why am I happy," is reading a newspaper dated 1898.

RHEINGOLD

As with Budweiser, Rheingold has been a popular and several-times-used brand name. While Rheingold is now the registered property of Rheingold Breweries, Inc. of Brooklyn, N. Y., there have been at least two other "Rheingolds" and one "Rhinegold" through the years.

Weisbrod & Hess was a small Philadelphia brewery founded in 1882 by George Weisbrod and Christian Hess. Obviously inspired by the masters, the firm's two brands were Rheingold and Shakespeare. Shown here is a beautiful 1908 brewery-scene calendar, a post-Prohibition tray and a bottle from the company's "Oriental Bottling Department."

Rhinegold Lager, as evidenced here by the pre-Prohibition bottle, was brewed early-on by Harry F. Bowler, of Amsterdam, N. Y. Originally founded as an ale brewery in 1889, Bowler added lager in 1891 and it's conceivable that he used the name Rhinegold from that date. T. Fitzgerald was most likely the local bottler of Bowler's beer and ale.

The longest-lived "other Rheingold" was a product of Chicago's United States Brewing Co. This company was formed in 1889 through the merger of three old-time Chicago breweries, Bartholomae & Leicht, Michael Brand, and Ernst Brothers. The United States Brewing Co. is known to have definitely made use of the Rheingold trade name well prior to Prohibition. In fact, the company even used a "Rheingold Girl" to promote its beer. In business until 1956, United States Brewing Co. included "The Story of Rheingold Beer" on its cans for many years. The close-up shown here is the reverse of the can pictured in the bottle and tray display.

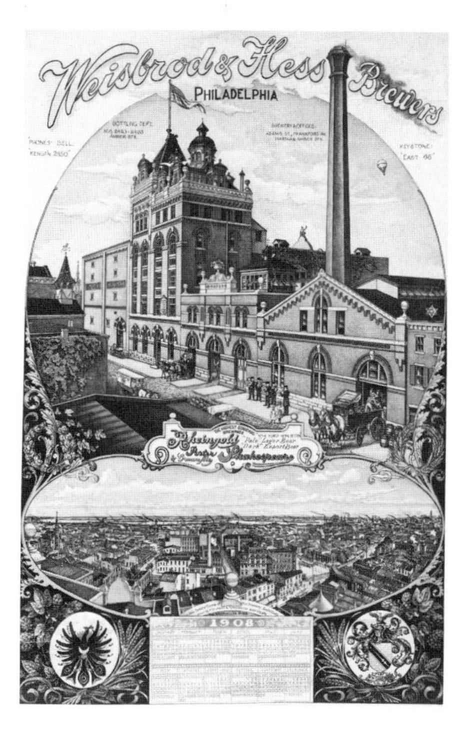

RHEINGOLD BREWERIES
36 FORREST STREET
BROOKLYN, NEW YORK

Today when one thinks of Rheingold, Brooklyn's Rheingold Breweries comes to mind. Founded in 1855 by Samuel Liebmann, the company was known as S. Liebmann's Sons Brewing Co. for most of its pre-Prohibition years. In 1924 when Obermeyer and Liebmann (another old Brooklyn brewing establishment, founded in 1868) was merged with the company, the name was changed to Liebmann Breweries. Forty years later, one last change was made, to the present-day name of Rheingold Breweries.

A pre-Prohibition embossed bottle and some early Rheingold cans. Rheingold was first put into cans in 1936, and the first design is seen on the can shown atop the pyramid to the right. Of much greater interest, however, are the Miss Rheingold cans to the left.

No one living in or around New York City in the 1940's, 1950's or early 1960's can forget the annual Miss Rheingold contest. It really was fun to see who would win, and to stuff the ballot box with votes for your own favorite! Ballot boxes were everywhere—taverns, delicatessens and grocery stores, and gave Rheingold great publicity.

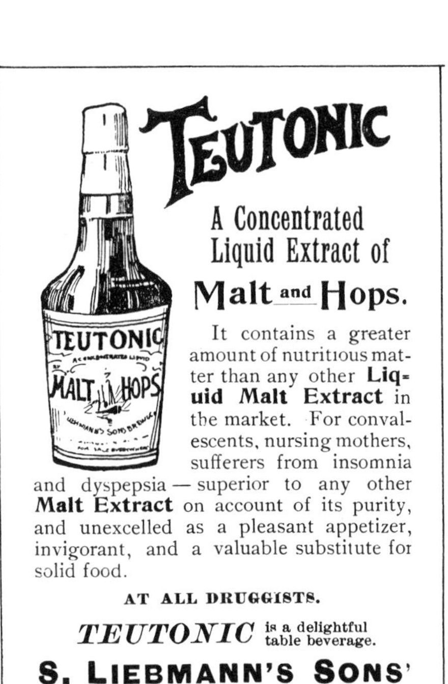

Though naturally best known for Rheingold, Liebmann also produced Teutonic Malt Extract for many years. Sold through drug stores, Teutonic is shown here as it appeared in a small 1895 magazine ad.

Pictured here on the can labels are the six contestants from 1957. Ironically, although she's now a movie star of some renown, Diane Baker (top can) did not win the contest. Miss Rheingold of 1957 was Margie McNally (bottom row, right).

MISS RHEINGOLD PHOTO QUIZ

How many Miss Rheingolds can you name?

1 2 3 4

5 6 7 8

9 10 11 12

13 14 15 16

17 18 19 20

(Answers on other side)

Twenty Miss Rheingolds, from 1940 to 1959. The contest ran from 1940 to 1965, when it was discontinued because of waning interest on the part of the public. The first and last Miss Rheingolds were selected by the company, but all other winners were elected on the basis of votes received. The very first Miss Rheingold was Jinx Falkenburg, who appeared in a number of 1940's movies, but who is much better remembered as the distaff side of the famous Tex & Jinx radio show.

PHOTO QUIZ ANSWERS

1. Jinx Falkenburg
2. Ruth Ownbey
3. Nancy Drake
4. Sonia Gover
5. Jane House
6. Pat Boyd
7. Rita Daigle
8. Michaele Fallon
9. Pat Quinlan
10. Pat McElroy
11. Pat Burrage
12. Elise Gammon
13. Anne Hogan
14. Mary Austin
15. Adrienne Garrett
16. Nancy Woodruff
17. Hillie Merritt
18. Margie McNally
19. Madelyn Darrow
20. Robbin Bain

JACOB RUPPERT
1639 THIRD AVENUE
NEW YORK, NEW YORK

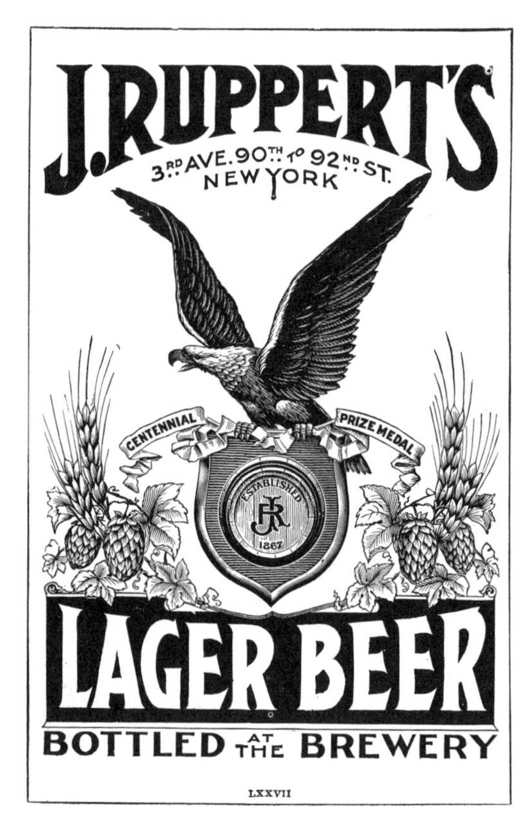

Ruppert must be considered to be one of the more famous of all beer/brewing names. Fritz Ruppert was a native of Bavaria who emigrated to America in 1835. After several years in the grocery business he established a malting company and, in 1850, bought a brewery. He called it the Turtle Bay Brewery due to its location in the part of the city known by that name.

Jacob Ruppert, Sr., was born in New York City in 1842. He naturally worked in his father's brewery and, in fact, was virtually manager by 1863. In 1867, however, he decided to strike out on his own, building a plant on Third Avenue and 92nd Street. The Third Avenue brewery prospered from the start and by 1916 Ruppert was the only non-national brewer to be in the 1,000,000-barrel-per-year category. Ruppert was also involved in ownership of quite a few other companies and was a well-known sportsman.

Actually, however, it was his son, Jacob, Jr., who carried the Ruppert name to even greater heights. First, he continued the great success of the brewery. Second, he served with considerable distinction during World War I, attaining the rank of Colonel. Third, because of his love for baseball he purchased the New York Yankees and rode them, with Ruth, Gehrig & Co., to national fame.

Those were the glory years (the Colonel himself died in 1939), and by the 1960's they were pretty much a thing of the past. The Third Avenue facilities were considered outmoded, and the company attempted to find an adequate site outside of New York City (and closer to New England—ironically the largest market for "New York's Famous Beer"). In fact, Ruppert came very close to constructing a $28,000,000 facility on a 96-acre plot in Carmel, N. Y. The problem of an adequate water supply, however, could not be resolved and the proposal eventually fell through.

In 1965, the famous old Third Avenue plant brewed its last batch of beer. The plant was closed and the Ruppert brand name sold to Rheingold. Rheingold has continued the use of Ruppert and Knickerbocker. In fact, the can still shows Jacob Ruppert, Inc. as the brewery. The Ruppert plant itself was, sadly, demolished in 1969 as part of an urban renewal project.

Needless to say, Ruppert used a number of different can designs over their 30-year canning period. I believe the can on the middle row, far right, in the three-tier set-up is the oldest, along with the bock beer version of the same can design (bottom row, extreme left of same three-tier set-up). These two look the oldest and also are the only ones that include instructions on opening. The two top cans in the same three-tier display were both for export. Notice the statement "Withdrawn Free of Internal Revenue Tax for Exportation."

The three cans directly to the left of the bottles are from Ruppert's Norfolk, Virginia, plant. This brewery was owned by Ruppert from 1948 to 1953. Brands included Red Fox as well as traditional Ruppert labels. The plant has been operating since 1954 as the Southern home of Champale.

The two labels are interesting. The top one is from the New York Bottling Co. of 514-520 W. 36th Street. Prior to the turn-of-the-century they were an independent

bottler used by Ruppert. The bottom label was used some years later, when the company was doing its own bottling right at the brewery.

The Baltimore branch bottle is of particular interest. Notice that the manager's name, Henry L. Lemkuhl, was embossed right on the bottle. Considering the expense of embossing, someone (either Ruppert or Lemkuhl himself, depending on who paid for the bottles) certainly had a lot of confidence in Mr. Lemkuhl.

ADAM SCHEIDT BREWING COMPANY

151 WEST MARSHALL STREET
NORRISTOWN, PENNSYLVANIA

In the year 1870 Moeshlin Brothers started a small brewery in Norristown. Several years later, in the late 1870's, C. & A. Scheidt bought the plant from Brothers. Small it was; the 1879 "List of Brewers in the United States" notes, "Scheidt, 699 barrels sold." The other brewer in town, A. R. Cox, although also extremely small, sold over three times as many barrels (2,228) that year. In 1884 Adam Scheidt became sole proprietor, and six years later the firm was incorporated as the Adam Scheidt Brewing Co.

Scheidt bounced back after Repeal, brewing Valley Forge, Rams Head, Scheidt's, and Prior. In 1955 the company was purchased by C. Schmidt & Sons of Philadelphia. It continued to operate under the Scheidt name until 1960. The name of the company was then changed to the Valley Forge Brewing Co. In 1965 it was changed again, this time to C. Schmidt & Sons, under which listing it continues in operation today.

Although the Washington's Headquarters rectangular tray has the look of a pre-Prohibition tray, it was produced during the 1930's. Continuing with the same very appropriate theme to promote their Valley Forge brand, Scheidt used the round Washington's Headquarters tray during the 1940's. The tray to the left is obviously newer–and also obviously uglier. It was used during the 1950's.

Looking very closely on the left side of the rectangular tray, one can read, "Philadelphia Branch, 2815-29 Ridge Avenue." Ironically, Scheidt is now a branch of a Philadelphia brewery.

Scheidt cans are fairly easy to date due to the three changes in name. Cans reading Adam Scheidt are from 1935-1959; those of Valley Forge Brewing Co. are from 1960-1964; C. Schmidt & Sons cans are from 1965 on. All the cans shown on this page are from the Adam Scheidt days (i.e., 1935-1959), the oldest being the three in the right foreground. The very unusual picture-of-the-plant design is from the back of the Rams Head Ale can that is the leftmost can of these three.

JOSEPH SCHLITZ BREWING COMPANY

235 W. GALENA STREET
MILWAUKEE, WISCONSIN

The company that produces "The Beer That Made Milwaukee Famous" dates back to 1849. In that year August Krug, a German immigrant, erected a small brewhouse on Chestnut Street. A year later members of both families that have been associated with Schlitz ever since joined the firm. First, August Krug hired a 20-year-old bookkeeper by the name of Joseph Schlitz. Secondly, he brought his eight-year-old nephew, August Uihlein, over to Milwaukee from Germany to live with him.

Krug died in 1856 and Joseph Schlitz assumed management of the brewery. Some time later he also married Krug's widow. In 1874 the name of the company was changed to the Jos. Schlitz Brewing Co. Ironically, a year after this name-change, Schlitz died. On an ocean voyage to visit his native Germany, the steamer "Schiller" sank in the Irish Sea and Schlitz and his wife were drowned. He was but 44 years of age. Management of the brewery passed to August Uihlein and his three brothers, Henry, Alfred and Edward. Today, 99 years later, Schlitz is still managed by a Uihlein. Robert A. Uihlein, Jr., present President and Chairman of the Board of Schlitz, is a grandson of August.

Now the second largest brewing company in the world, Schlitz has had many milestones over the years. The most interesting and perhaps most important dates back to 1871, the year of the Chicago Fire. Because of this disaster most of the "Windy City" was left without water. Schlitz helped out by shipping down barrels and barrels of beer, thus taking its first step away from purely local marketing to its present world-wide distribution. Out of this fire also came one of the most famous of all slogans—"The Beer That Made Milwaukee Famous."

Jos. Schlitz

Up until 1949 Schlitz was a one-plant (in Milwaukee, naturally) brewing company. In that year, however, it started down the road to its present multi-plant system when it purchased this Brooklyn plant.

While many, many improvements have been made over the years this facility traces its roots back to 1866, when Leonhard Eppig and Hubert Fischer erected the Eppig and Fischer Brewery. In 1876 Fischer left to form his own brewery in Hartford, Conn., and the company became the Leonhard Eppig Brewing Co.

Eppig did not re-open after Prohibition, but the facility was reactivated by George Ehret, and "Ehret's Extra" was produced here until 1948. Schlitz moved in a year later and used the brewery as its Northeast production facility until early 1973. In the years since 1949, however, Schlitz had built a number of brand-new and very efficient breweries, including a huge facility in Winston-Salem, N. C., opened in 1969. In Winston-Salem rail facilities, land for expansion, and cheaper labor were all prevalent. In Brooklyn they were not. By the early 1970's it was cheaper to brew in North Carolina and ship the beer 500 miles to New York than it was to brew in New York itself. Result, the closing of another facility. The above photograph was taken in April, 1973, one month after the last batch of beer brewed here had been shipped.

A dramatic shot–a tray with a post-Prohibition labeled bottle posed above rows of turn-of-the-century Schlitz agency bottles. Schlitz became so synonymous with Milwaukee that often "Schlitz" was left off and "Milwaukee Lager" embossed. The entire right-hand row consists of Manchester, N. H. bottles. Schlitz either had a lot of competing agents in Manchester, or else had a terrible turnover problem. The Milwaukee Club cans are from the late 1930's or very early 1940's, as this was a Schlitz secondary brand that ceased to be marketed during World War II.

Some impressive early signs—plus a dictionary. The dictionary was entitled the *Schlitz Milwaukee Dictionary,* and was copyrighted in 1899. A copy would "be mailed to any address on receipt of five cents in postage stamps to cover cost of mailing." The words of the Preface alone are worth 5¢ today: "The time has gone by when to spell incorrectly was thought to be a matter of not much moment. In these go-ahead days, these days of educational advantages, poor spelling is regarded as almost synonymous with ignorance. In order to obtain any position in an office—as bookkeeper, amanuensis, typewriter, and so forth—it is now an absolute necessity that one's orthography should be perfect. And in one's private correspondence, too, correct spelling is deemed to be quite as important as is good writing.

Such being the case, the possession of a carefully compiled Dictionary, one which is both handy to carry and handy to refer to, is a something of the greatest possible help to everybody."

While there is a small advertising message on the bottom of each page, Schlitz did not include promotional statements in the factual area, with one exception. In the back of the dictionary there is an interesting section that includes all kinds of factual information. Along with Legal Holidays ("Every Saturday after 12 o'clock noon is a legal holiday in N. Y., N. J., Penn., and Md., and the city of N. O., and June 1 to Sept. 30 in Newcastle county, Delaware"), Recipes for Housekeepers ("For Bites of Mad Dogs—Apply caustic potash at once to the wound and give enough whiskey to cause sleep") and Hints on Etiquette ("Permission must be obtained when a gentleman is presented to a lady"), there is also a list of unusual facts. Here, Schlitz did succumb to the temptation, and included "The products of the Jos. Schlitz Co.'s Breweries are to be found on sale in every civilized country in the World" in the list, immediately before the fact that "When a dog barks at night in Japan the owner is arrested and sentenced to work a year for the neighbors that were disturbed."

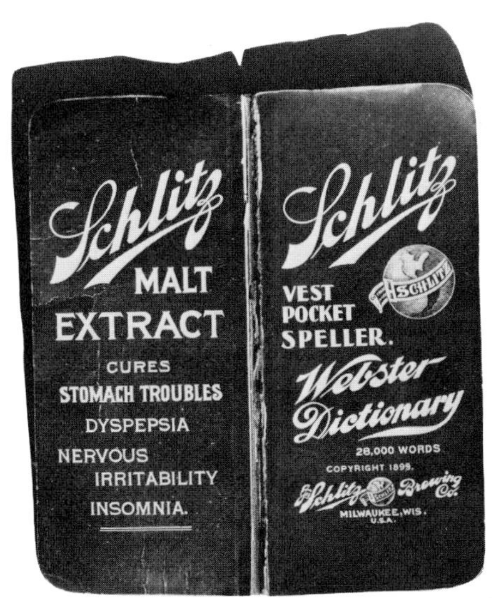

SCHWARZENBACH BREWING COMPANY
GERMANIA & GALETON, PENNSYLVANIA
HORNELL, NEW YORK

In 1857, when Germania, Pa., was still a backwoods community, Joseph Schwarzenbach founded a brewery there. For years the water for the brewery was carried in pails from a nearby brook and it was necessary to take a four-day trip by ox-team to the nearest railroad station to get supplies of hops. The brewery prospered, however, and became a little more modern year by year.

Joseph died in 1892 and the brewery was taken over by his three sons, Roland, Herman and James. They continued the operation in Germania and also established, in 1895, a second Schwarzenbach Brewing Co., in Hornellsville (now known as Hornell), N. Y. In 1902, they also built a plant in Galeton, Pa., and closed the old brewery at Germania. Both plants produced cereal beverages and soft drinks at the start of Prohibition. Howevery, when Repeal finally came, only the Hornell facility opened up—as the Hornell Brewing Co.

Brands were Schwarzenbrau, Hornell, Old Ranger and KDK. In 1960 the company was taken over by the Benjamin Hertzberg Metropolis combine. It didn't, however, last long as part of that brewery chain; the plant was closed in 1964.

BREWERY AT GERMANIA, PENNSYLVANIA, 1880.

Both trays are from the Hornell Brewing Co.; therefore, both were produced after Repeal. In addition to their own brands, Hornell also produced private label brands. What is interesting to note here is that they brewed for both A & P (Tudor) and Grand Union (Gilt Edge), two strong supermarket competitors.

It was in 1906 that Hornellsville shortened its name to Hornell. Hence, the two middle blob-top bottles were produced sometime between 1895 and 1905, while the clear crown quart in the rear is from the 1906 to 1918 period.

PETER SCHOENHOFEN BREWING COMPANY

CANALPORT AVENUE &
18th STREET
CHICAGO, ILLINOIS

A very sizeable Chicago brewery, this company was founded in the very early 1860's by Peter Schoenhofen. Schoenhofen was a Prussian by birth, and had worked in the related fields of distilling and cider manufacturing prior to his involvement in brewing. From the time he opened the brewery until his death in 1893, sales grew at a very rapid rate—from 600 barrels the first year to approximately 180,000 barrels in the year of his demise. After his death, the brewery was very ably managed by Schoenhofen's son-in-law, Joseph Theurer.

After Repeal the brewery, now called the Schoenhofen-Edelweiss Co. (Edelweiss had always been the company's prime brand) reopened at the same 18th Street address. It operated independently for 20 years, until 1954, when it became a division of Drewry's, Ltd., U.S.A.

In that same year the Schoenhofen-Edelweiss Co. was "moved" to a different plant, at 2132 South Laflin Street, although Drewry's kept the old 18th Street facility in operation as "Drewry's Plant #2." The Schoenhofen-Edelweiss name is not listed in directories after 1962, but the original Peter Schoenhofen brewery operated, always as Drewry's #2 plant, up until 1971.

A tray, copyrighted in 1913 by the brewery, and some examples of cans used by Schoenhofen-Edelweiss.

A Floor in One of the Cellars

The Coopershop

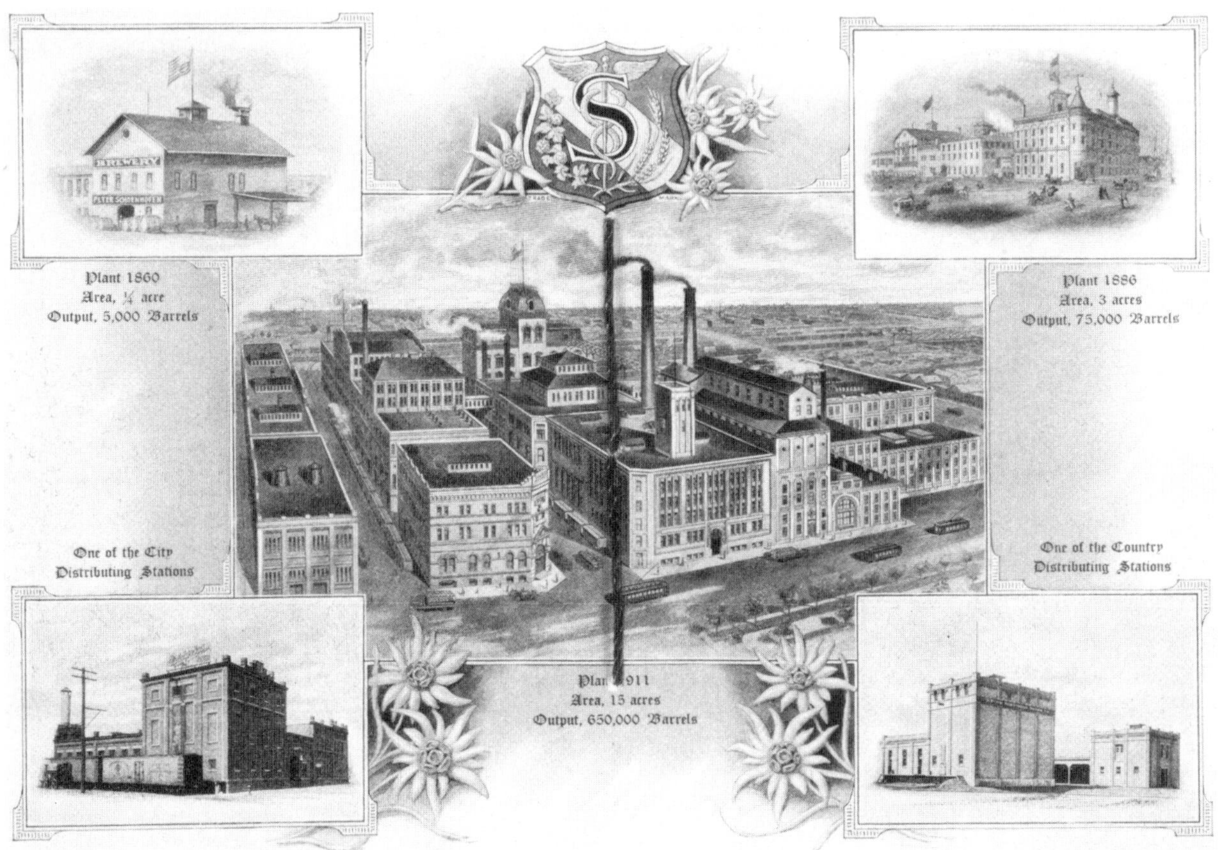

The Labeling Department

Prior to Prohibition, if you took a
presented with a little booklet as
a souvenir of the visit. These booklets
were generally beautifully produced
and printed, and are today extremely
brewery tour you were most likely
collectible. Sixty years ago a "tour" of
the Peter Schoenhofen Brewery, as
depicted in pages from the company's
souvenir booklet (copyright 1911), would
have included these highlights.

The Filling and Crowning Department

SPOETZL BREWERY
BREWERY STREET
SHINER, TEXAS

In South Central Texas, deep in the heart of cattle and cotton country, stands one of America's smallest operating breweries, Spoetzl.

Originally founded in 1909 by a group of Shiner businessmen, it was not an immediate success. In fact, in its first six years of existence the company nearly went bankrupt twice. As a result, in 1914 the owners began looking for someone with brewing experience to take over. They found Kosmas Spoetzl, who was Bavarian by birth but who had traveled considerably, and had brewing experience in such disparate cities as Cairo, Egypt, and San Antonio, Texas.

Brewing Old World Bavarian Beer from a recipe handed down for generations in his family, Spoetzl turned the brewery into a successful venture. When Prohibition set in, Spoetzl remained open, producing near beer and, later, ice. Much of the ice was undoubtedly sold to local residents who were lagering their own home-brew!

Kosmas Spoetzl died in June of 1950, at the age of 77, and his daughter, Cecelie, became sole owner of the brewery. For 16 years she operated the company—the only proprietress of a brewery in the United States during that time—until she sold it in 1966. Spoetzl has since become incorporated, and is still very much alive and brewing today.

The two very unusual dogs-playing-poker scenes have been used by Spoetzl on their calendars for the last 15 years. The can, however, is much more recent. It was not until September of 1970, 35 years after Krueger sold the first can of beer, that Spoetzl started packaging its beer in cans. This fact, however, is in no way presented to belittle Spoetzl—but merely to show that major packaging changes are something that a small brewer has to weigh very carefully. Canning (or any other type of new machinery) equipment does not come cheaply. Thus, it is interesting to note Spoetzl's time-table of packaging changes over the years:

1909—Wooden Kegs
1916—Glass Bottles
1947—Aluminum Kegs
1958—No Return Bottles
1964—Party Kegs
1970—Cans

STANDARD-ROCHESTER BREWING COMPANY

770 EMERSON STREET
ROCHESTER, NEW YORK

In 1956 the Standard Brewing Co. (Standard, Ox Head and Ox Cart brand names) and the Rochester Brewing Co. (Old Topper) consolidated to form the Standard-Rochester Brewing Co. Both of these companies had their origins prior to Prohibition (Standard in 1889 and Rochester in the 1870's). When they merged, the former Standard plant at 436 Lake Street was shut down and all brewing took place at Rochester's 770 Emerson Street facility. For 14 years the consolidated company survived but in 1970 the end came.

Auctioneers, Auction Sale 12

BREWERY AUCTION
2 DAY SALE

By order of the Bankruptcy Court in the matter of Standard Rochester Brewing Co., Inc., Bankrupt. 2 day sale. We will sell at Public Auction on Tuesday, October 13, 1970 at 10:30 A.M. and Wednesday, October 14, 1970 at 10:30 A.M. on the premise at 770 Emerson Street, Rochester, New York. Complete brewery production equipment, modern bottle and can lines, brew house equipment, mash, fermenting and storage v[illegible] and tanks, materials hand[illegible] warehouse equipment [illegible] systems, Fleet of tr[illegible] trucks, [illegible] trucks, [illegible] ment [illegible] -Cut- [illegible] oledo 999 [illegible] ale etc. power [illegible] oists and falls; 5- [illegible] ion and floor jacks; 6- [illegible] 2,500 lb. hydraulic lift jacks; 59,000 bottles; 4,000 aluminum kegs; 500- stainless kegs 2,000 wood pallets. Large quantity woodworking & plumbing machinery, restaurant equipment, laboratory equipment supplies scrap etc. 9.7 acres of prime light industrial real estate land & buildings will be offered for sale. 6-"C" licenses will be offered for sale. All sales subject to confirmation by the Hon. Austin J. Donovan, Referee in Bankruptcy, E. Charles Larry, Trustee, Burns, Suter & Doyle attys. for Trustee. Premises open for inspection, Monday, October 12, 1970, 10:00 A.M.-4:00 P.M. Tuesday, October 13, 9:00 A.M. until auction & Wednesday, October 14, 9:00 A.M. until auction. For complete information phone David Weinstein, I. SHOOLMAN & ASSOCIATES, Inc., Auctioneers, Phone 442-8720.

Three prior-to-consolidation signs, back when Standard and Rochester were competitors instead of cohorts.

A real assortment of Standard and Rochester containers used through the years, including a series of dated 1890's Rochester Boston branch bottles. Unfortunately, Standard-Rochester invested heavily in the gallon can concept in the mid-1960's and it flopped for them.

STEGMAIER BREWING COMPANY

152 E. MARKET STREET
WILKES-BARRE,
PENNSYLVANIA

This brewery was founded by Charles Stegmaier, born in Württemberg, Germany in 1821. He practiced brewing from the age of 15 in his homeland and then, in 1849, came to America. He worked for several Philadelphia breweries, and then joined in partnership with John Reichard in Wilkes-Barre. This was 1851, and Reichard and Stegmaier are credited with being the first to brew lager in northeastern Pennsylvania.

In 1857, Stegmaier went out on his own, establishing a bottling company. Shortly after this, he entered into partnership again, this time with George C. Baer. They built a small brewery on South Canal Street, and did quite well until the Panic of 1873 forced them to suspend operation. Stegmaier went into the hotel business and then, in 1875, rented the Bowkley Brewery on North River Street and went back to brewing. With his son now in partnership with him, Stegmaier stayed in business in this manner until 1880 when he purchased the former Baer & Stegmaier facility. Back owning his own brewery again, Stegmaier proved a very capable businessman, and the business grew considerably. By the turn-of-the-century the Stegmaier Brewing Co. was the largest brewery in the state of Pennsylvania outside of Philadelphia and Allegheny (Pittsburgh) counties.

Stegmaier is still very much in business, brewing Stegmaier Ale, Porter, Gold Medal Beer and Stallion XII Malt Liquor. The name "Gold Medal" comes from the many European exposition awards won by Stegmaier between 1910 and 1913.

Stegmaier has issued a number of different trays over the years. The two on the upper right are very much pre-Prohibition while the three on the bottom row are very much post-Prohibition. The plant scene in the upper left is a fooler, though. You'd expect it to be 1930's or early 1940's, or maybe even late pre-Prohibition, right after the awards were won. But, according to company records, it was issued in 1952. The two brown spout cans on the right were used from approximately 1940-1944, while the white can in the center and the white and brown can on the left were both put in use at about the same time, 1944.

STROH BREWERY COMPANY

331 GRATIOT AVENUE
DETROIT, MICHIGAN

Bernard Stroh was born in Germany in 1822. While in his 20's he left Germany for the "New World." At first his goal was Brazil, where a sizeable German and Austrian community had developed. However, he was not long in South America before he realized that there were greater opportunities in the United States. While only intending to pass through Detroit, he took an immediate liking to the city and decided to take up residence there.

Having learned the brewing trade in Germany, he set about to found his own brewery in Detroit, which he did in 1850. Originally called the Lion Brewery, the company was a success from the start. By the time Bernard Stroh passed away, in 1882, the Lion Brewery was the largest in Michigan. Under his two oldest sons, the business continued to grow. In fact, when Prohibition officially began in Michigan on May 1, 1918, Stroh was producing over 300,000 barrels of beer per year.

During Prohibition Stroh stayed in business, producing ginger ale, ice, malt extract and ice cream (a product it still produces today). With Repeal in 1933, it was naturally back to beer, and it has been ever since. Stroh is now one of the ten largest breweries in America.

Five pre-Prohibition Stroh items. The tray on the lower right is unusual in that it is two-piece construction—enamel middle with a brass rim. Presumably it is of the same vintage (probably late 1890's) as the boy-with-case tray, since both state, "Annual Capacity 500,000 Barrels" and "Highest Award Medal, World's Fair 1893."

In April of 1964 Goebel, another old-time Detroit brewery, was acquired by Stroh. Shown are some older post-Prohibition items from the manufacturer of "Mello-ized" beer.

WEST BEND LITHIA COMPANY

459 N. MAIN STREET
WEST BEND, WISCONSIN

West Bend traces its beginnings back to two of the pioneer breweries of Wisconsin, one founded in 1849 by B. Goetter and the other founded in 1850 by Christopher Eckstein. After many years of separate operations these two small breweries were merged together in 1883, forming the West Bend Brewing Co.

In 1911 the brewery was sold to a group of businessmen headed by Martin F. Walter, of the famous Walter brewing family. Members of this family founded and have operated the George Walter Brewing Co. of Appleton, Wis.; the Walter Brewing Co. of Eau Claire, Wis.; the Walter Brewing Co. of Menasha, Wis.; and the Walter Brewing Co. of Pueblo, Colorado, as well as operating West Bend. An impressive list for any family!

West Bend continued in operation, under Charles W. Walter, Jr., until 1972. It ceased operations as of October 15th of that year.

A few interesting West Bend items. The aluminum tray on the right was purchased just prior to W.W. II, while the Old Timer's tray was originally ordered around 1949 or 1950, and subsequently re-ordered several times after that. Supposedly the very happy looking gentleman on the tray is among the men seated at the table on the Old Timer's Lager label, although it is hard to tell which one he is.

Milwaukee Valley and Black Pride were two short-lived sub-brands put out by West Bend. Milwaukee Valley cans were purchased but once (250,000 quantity) about 1959 and sold only in California. Black Pride was sold only during 1969 and 1970.

What is Lithia and how did it end up in the West Bend name? The answer is complex but sensible. The word "Lithia" is a derivative of "Lithium Carbonate" and is a type of mineral water. West Bend used this special water and even considered bottling it separately at one point. During Prohibition, when the word "Brewing" was supposed to be dropped from the company's name, it substituted Lithia. And it remained West Bend Lithia Company right up until the end.

WEST END BREWING COMPANY

811 EDWARD STREET
UTICA, NEW YORK

This brewery's roots go back to 1856 when Charles Bierbauer started operations on the present site. Bierbauer manufactured lager beer until his death in 1885. In that year the business was merged into the Columbia Brewing Co. In 1887 it became the West End Brewing Co., under the supervision of F. X. Matt. Matt, who was born in Germany in 1859, came to this country in 1878, and had gained valuable experience at the Duke of Boden Brewery in Rothaus, Germany, the Bierbauer Brewery in Utica and Louis Bierbauer's Brewery in Canajoharie, N. Y. (Louis was Charles' brother). In those early years West Bend was the smallest of nine Utica breweries, producing but 4,000 barrels of beer a year.

Through the years the company, always under the control of the Matt family (in fact, F. X. Matt served as President up through 1950), has increased in size and sales. Brands are Utica Club (a name adopted during Prohibition, when the company turned to soft drinks, malt tonic, extracts, etc.) Beer & Ale, Matt's Premium Beer and Fort Schuyler Beer (a brand name acquired by purchase of the Utica Brewing Co. in the early 1960's).

The tray on the left is known as the "Miss Columbia Tray" and is pre-Prohibition. The plant scene tray, however, is a post-Prohibition item and was used through the late 1940's. Of the two bottles, the one on the left is pre-Prohibition, while the smaller bottle on the right is a 6½ oz. soda bottle. West End produced a wide line of soft drinks under the Utica Club name beginning in 1919 and continued production of soft drinks after Repeal until materials became very scarce during World War II.

The two signs are relatively new. The Matt's is from the mid-1950's and the Old English Ale sign was used up until 1960 when West End discontinued Old English as a package brand (it is still available today, but on a very limited basis and only as a draft product).

The spout can in the left-front (used from 1938 to 1940) is a beautiful blue. On the back of the can is the message, "When Empty Throw The Can Away." Fortunately, somewhere along the way, somebody disobeyed this "order."

D. G. YUENGLING & SON

501 MAHANTONGO STREET
POTTSVILLE, PENNSYLVANIA

This is America's oldest operating brewery. Others may imply that they are, but none go back as far as Yuengling's 1829 date.

The company's founder, David G. Yuengling, was born in Germany in 1806 and came to America in 1828. After a year spent in other parts of Pennsylvania, he settled in Pottsville in 1829. In 1873 Yuengling's son, Frederick G., came into the firm and the name of the brewery was changed to its present D. G. Yuengling & Son.

In 1873 another son, David G. Yuengling, Jr., established an ale brewery in New York City, at 5th Avenue and 128th Street. Several years later he purchased a second brewery, for lager, at 10th Avenue and 128th Street. He operated both breweries under the Yuengling name for many years. However, these operations were generally run at a loss and continued only because of financial support from Pottsville. Finally this support was cut off and in 1884, therefore, the 5th Avenue facility was closed. The 10th Avenue facility operated until 1897, when it was sold to John F. Betz, and became known as the Manhattan Brewery.

Yuengling also operated a short-lived, late-19th century brewery in Richmond, Va. (Betz, Yuengling and Beyer), and an equally short-lived company in Hudson, N. Y. The Hudson address was 2nd and State Streets, and it lasted but a few years in the very early 1900's.

The Pottsville plant stayed open all through Prohibition, producing cereal beverages. Today, it is justly proud of the fact that it is the OLDEST BREWERY IN AMERICA and is owned and operated by AMERICA'S OLDEST BREWING FAMILY. Yuengling's president today is Richard L. Yuengling, a great-grandson of Yuengling's founder.

YUENGLING BREWERY, POTTSVILLE, PENNSYLVANIA, 1844.

Over a hundred years of Yuengling collectibles are shown here. It is the brewery's belief that the tall black and gold sign is from the 1860 era, and is certainly from well before the turn-of-the-century. It is made of thick glass adhered to a block of wood that's over one inch thick.

The bottle on the far left would indicate that Yuengling had a brewery in Rye Neck (Westchester County), N. Y. Records, however, do not support this and Rye Neck was probably a branch or depot for the New York City operation, as was Saratoga, N. Y.

All trays are from the period 1939 to 1948. The lower right tray was manufactured in 1939; the horsehead tray in 1941; the bottle and glass tray in 1946; and the eagle tray in 1948.

"Yuengling's Special" was a Prohibition cereal beverage put out by the company, and the small round sign in the upper left-hand corner is from around 1930.

Yuengling first packaged its beer in cans in the summer of 1938. The cans were spout-top and Yuengling's was written as on the tall black sign, with the letters connected together and the sweeping tail. The style was changed slightly in 1946 and the ale can and two "beer" cans show samples of this. Use of the two "Premium beer" spouts started in 1948.

GREAT BREWING CITIES

This type of case, with the hinged top, is the kind generally used before Prohibition. Fort Pitt was founded c. 1907 and remained in operation until 1957.

Introduction

When one thinks of American cities famous for their beer and their breweries, such large urban centers as Milwaukee, St. Louis, Brooklyn and Cincinnati immediately come to mind. However, size in itself did not guarantee that a community would be a major brewing center. Los Angeles, Houston, Denver and Washington, D. C.—all large cities—were never important with respect to malt liquor output.

What factors, then, did contribute to the growth of the brewing industry in certain cities? Generally, there were four reasons:

1) The presence of a large German-American citizenry from which skilled brewery workers were drawn and by which a beer-drinking tradition was established and upheld.

2) A reliable source of natural ice. This was an absolutely essential ingredient prior to the development of artificial refrigeration in the early 1880's, since lager beer has to age in a cool place. Without ice there was simply no way to provide the necessary degree of cold. This one requirement removed the entire South from serious beer production until the late nineteenth century. By then, the major brewery centers had long been established.

3) An adequate water supply. Most any water will suffice for brewing since it is so highly purified, but beer is made up of approximately 92% water. Without plenty of it no city could ever support many breweries.

4) Size of population. As explained, size alone has not guaranteed brewing importance, but it has been a contributing factor. Many cities in the Northeast and throughout Wisconsin, Minnesota and Iowa possessed all the other necessary ingredients for brewing success, but they lacked a sizeable enough home market base. Ironically, in later years, the contrary situation prevailed. Many breweries in large metropolitan centers did not feel compelled to look outside their own immediate area for customers and, therefore, failed to take proper steps toward regional and national distribution. Milwaukee brewers did not make this mistake. Schlitz, Pabst, Blatz, and Miller were large and ambitious operations with a comparatively small local market base. To support their ambitions, they reached beyond Milwaukee to the rest of the Midwest, and eventually, to the entire nation.

Eleven cities rose to the forefront of American brewing—Boston, Brooklyn, Buffalo, Chicago, Cincinnati, Detroit, Milwaukee, New York, Philadelphia, St. Louis and San Francisco. With the exception of Milwaukee, these cities claimed the greatest number of breweries at the turn-of-the-century, the heyday of American brewing history. Milwaukee, while never noted for its *number* of breweries, has certainly always been known for its *output*—and no account of major brewing cities would be complete without it.

Eight others certainly deserve "honorable mention." These, in order of their number of brewers at the turn-of-the-century, are: Baltimore, Newark, Cleveland, Louisville, Albany, Syracuse, Rochester and Pittsburgh.

Boston

Of the eleven major brewing cities, Boston is the most surprising entry. In fact, it seems inconceivable that Boston, known for its Irish and Italian traditions, was home to more breweries than Milwaukee all through the pre-Prohibition years. The number of breweries in operation for selected years is as follows: 1879—20, 1890—21, 1910—20, 1935—3, 1950—4, 1960—1, 1973—0.

During the late 1880's, due to a depression in Great Britain, many English financiers looked abroad to invest their money. For some reason American brewing stock caught their fancy. As a result, starting in 1888 in Philadelphia and New York, British investment syndicates sought to gain control of numerous companies. Boston was especially affected by this British "invasion." Four Massachusetts breweries, including Roessle (bottom row, extreme left) and Suffolk (top row, second from right) were consolidated to form the New England Breweries Company, Ltd., an English syndicate, in 1890.

Eleven years late, in 1901, probably as a defense mechanism against this English operation, ten additional Boston breweries banded together under the name of the Massachusetts Breweries Company. These included Habich & Co.'s Norfolk Brewery (middle row, three extreme left bottles), H. J. Pfaff (middle row, second from right), and both plants of the Wm. Smith & Sons Brewing Co. (top row, extreme right and second from left; middle row, third and fourth from right).

John Roessle (bottom row, extreme left) was the first brewery in New England to brew lager beer—in 1846.

Notice the three bottles with dates on them (all in the middle row). This dating practice seemed to be especially popular with New England breweries.

STAR BREWING COMPANY
STAR BREWING CO.
FINE ALES AND LAGER
BREWING CO
BOSTON
SMITH & ENGLE
REVERE
LAGER BEER
BREWERS &
BOTTLERS
BOSTON, MASS.
SUFFOLK
BREWING CO.
BOSTON
BOSTON
MASS

NORFOLK BREWERY
REGISTERED
1894
BOSTON, MASS.
NORFOLK BREWERY
HABICH & CO. BOSTON
REGISTERED
NORFOLK BREWERY
DEPOT
361 ATLANTIC AVE
BOSTON
C.F. BURKHARDT
BOSTON
MASS.
1896
Spring
Bottling
Dept
Boston, Mass
SMITH & ENGLE
ELMWOOD
SPRING BREWERY
ROX BURY
H. & J. PFAFF
BREWING CO.
Registered
Boston
1897

THE ROESSLE
BREWERY
BOSTON
REGISTERED
BOSTON, MASS.
COMMERCIAL'S
Old
India Pale
Ale
BOSTON, MASS.
HINCKEL BREWING CO
BOSTON
MASS.
SPARKLING
LAGER BEER

Brooklyn

Brooklyn is today, of course, one of the five boroughs of New York City. It was, however, an independent city until 1898, and has always been a major brewing center. Natives claim many more honors. The number of breweries has been on a steady decline since 1879: 1879—43, 1890—38, 1910—31, 1935—10, 1950—7, 1960—4, and 1973—3.

Of all the breweries represented in this illustration, only Piel Brothers (top row, fifth from left) is still in business. Piels also owns the rights to the Trommer's (bottom row, fourth from right) name and does market a small amount of Trommer's White Label. Trommer itself went out of business in 1950.

The four bottles on the left of the bottom row were all utilized by early soda bottlers who also packaged beer. All say, "Brooklyn Garden Lager Bier" or "Excelsior Bottled Lager Bier" on the reverse.

Notice the two bottles to the right of the tray—"Leonard Eppig's Germania Brewery" and "Joseph Eppig's Brewery." It is thought that these two gentlemen were brothers, and presumably, friendly competitors.

Old Dutch was solely a post-Prohibition brewery, operating from 1934 to 1950.

In addition to Piels, Brooklyn's two remaining breweries are the F. & M. Schaefer Brewing Co. and Rheingold Breweries.

BREWERY
CONSUMERS
WASHINGTON &
FRANKLIN AVES
BROOKLYN
BROOKLYN N.Y.
PIEL BRO'S
OTTO HUBER BREWERY
BROOKLYN N.Y.
BROOKLYN N.Y.
MELTZER BROS
BREWERY
BROOKLYN, N.Y.
CONGRESS
TABLE BEER
BROOKLYN
CONGRESS
BREWING CO. LTD.
169
MESEROLE ST.
BROOKLYN
N.Y.
THE FRANK BREWERY
BROOKLYN N.Y.

ADOLF SCHMIDT'S
KLOSTER BREWERY'S
BROOKLYN N.Y.
JOHN KISSEL & SON
BROOKLYN N.Y.
THE EASTERN BREWING CO.
TRADE MARK
REGISTERED
BUSHWICK AVE
BROOKLYN N.Y.
EURICH BREWING CO.
BROOKLYN N.Y.
REGISTERED
Old Dutch
LAGER BEER
ALE & PORTER
BREWERY
BROOKLYN N.Y.
LEONHARD EPPIG
GERMANIA BREW
BROOKLYN N.Y.
LAGER BIER
L. Michel
Brewing Co.

427
BROADWAY
BROOKLYN E.D.
5
GEORGE BOHLEN
358 & 360
HART ST.
BROOKLYN
367
JOSEPH FALLERT
BREWING CO.
BROOKLYN N.Y.
BROOKLYN, N.Y.
FULL PINT
PURE BEER
PROSPECT PARK BREWERY
BROOKLYN
BREWERY
BROOKLYN N.Y.
BROOKLYN N.Y.
256-268
MAUJER ST
BROOKLYN N.Y.
MALCOM BREWING CO.
BROOKLYN
N.Y.
REGISTERED

Buffalo

A city you probably would not expect to find in a list of major American brewing centers, Buffalo, nevertheless, has had a considerable number of breweries, including some fairly large ones. The number of breweries in operation for selected years is as follows: 1879—30, 1890—20, 1910—19, 1935—8, 1950—6, 1960—2, and 1973—0.

Both breweries represented by the two trays on the left of the top row of the illustration—Lang and Beck—go way back to the same year—1840.

While Gerhard Lang did not become involved with the company until 1862 (when he married the oldest daughter of the founder, Phillip Born), the brewery did not really prosper until he assumed control. By the turn-of-the-century Lang was a very sizeable 300,000-barrels-per-year-capacity plant. Gerhard Lang re-opened after Repeal but never did well. In fact, the brewery's listed capacity in 1941 was only 150,000 barrels, half the capacity it had 40 years earlier. The brewery closed in 1949.

Beck's, as with Lang's, was founded by someone other than the man who made it famous. The founder was Joseph Friedman, and it was not until 1855 that Magnus Beck became part owner of the company. Within five years, however, he was sole owner, and in 1865 he constructed a huge new plant, at 467 North Division Street. Bottling was begun by the brewery in 1882. As indicated by the post-Prohibition tray shown, Magnus Beck did operate successfully for a goodly number of years after Repeal—until 1956.

John Schusler (right-hand bottle, middle row) founded this venerable Buffalo brewery in 1859, and owned and operated it until 1896. In that year William Simon (who, it is believed, was Schusler's brewmaster) became proprietor, and, of course, the company name was changed. The William Simon Brewery was never that large, but it outlived every other Buffalo brewing company, many of which were considerably larger. In early 1973, however, the last bottles and cans of Simon Pure rolled out of the old Emslie Street plant, and another *once* major brewing city was left without a single brewery.

GERHARD LANG BREW
Lang's
Highest Quality
BEER AND ALE

BUFFALO'S BEST BEER
Since 1855
Beck's
BUFFALO'S BEST
Beer
Naturally Smooth

Schreiber's
MANRU
BEER-ALE
Thoroughly Delicious

Phoenix
BEER
MOFFATS
ALE
PILSENER
THE INTERNATIONAL BREWING CO BUFFALO, N.Y.
BREWING CO.
BUFFALO, N.Y.
Iroquois
INDIAN HEAD
BEER & ALE
BUFFALO, N.Y.

GERMAN-AMERICAN BREWING CO BUFFALO, N.Y.
Stein's
ALE
BREWING CO
BUFFALO, N.Y.
SIMON PURE
The Seal of Good Taste
ALE - BEER
EXTRA PALE ALE
GERHARD LANG BOTTLING WORKS BUFFALO, N.Y.
BUFFALO N.Y.
THE MAGNUS BECK BUFFALO, N.Y. BOTTLING WORKS

Chicago

Although always a major brewing city, Chicago never really became the dominant force in American brewing that one might expect of the nation's second largest city and Midwestern capital. Two possible reasons stand out—one, many Chicago breweries never fully recovered from the disastrous Fire of 1871, and, two, neighboring Milwaukee's several large and ambitious breweries early aimed their sights at Chicago's much larger market, and succeeded in capturing a large portion of it. Nevertheless, Chicago breweries have been many in number: 1879—20, 1890—41, 1910—54, 1935—27, 1950—18, 1960—10, and 1973—1.

Illustrated are several Chicago brewery trays and spout quart cans. Yusay Pilsen was the primary brand name of The Pilsen Brewing Co., founded c. 1907 and remaining in business until April of 1963. The second tray has a pre-Prohibition look to it, but it is from a brewery, Ambrosia, that operated only after Repeal—from 1938 to 1959. Canadian Ace was also a post-Prohibition company, in business from 1948 to 1968.

Best, on the other hand, has a history beginning in 1885, although it was not known as the Best Brewing Co. until 1890. After Prohibition, it re-opened and brewed Best, as well as Hapsburg and Embassy Club, until 1963.

Koenig-Brau, was a product of the Prima-Bismark Brewing Co. Prima and Bismark both opened in 1933, and were merged in 1941. Prima was first located at 825 Blackhawk Street and, in 1938, moved to 4160-76 South Union Ave. Its own plant was closed in the merger with Bismark, and all brewing operations were conducted at the latter's plant at 2762 Archer Avenue until 1952 when the merged firm joined the ranks of the brewery deceased.

Keeley (can on far right) is an unusual name for a brewer; Michael Keeley was an Irishman, and had to fight hard to achieve success. The author of *One Hundred Years of Brewing,* published in 1903, had this to say at the time: "The founder of the extensive business represented by this company was one of several Irishmen among Chicago brewers, who, through perseverance and a display of Irish grit and intelligence, achieved great success in a field which previously had been almost monopolized by the Germans." Keeley was born in Carlow, Ireland, and came to America as a teenager, residing in Chicago from 1850 until his death in December of 1888. As a post-Prohibition brewery, re-opening in 1934, the company managed to operate until 1952.

Maintaining the heritage of Chicago's brewing glory past is Old Chicago Lager, produced by the windy city's sole remaining brewery, the revived Peter Hand Brewing Co.

A very beautiful label from the strictly post-Prohibition Atlantic Brewing Co. Atlantic opened in 1933 and grew to the point where, at one time, it operated two Chicago plants and owned the Bohemian Breweries of Spokane, Washington, and Boise, Idaho. As with so many other breweries, however, bad times set in during the 1950's and 60's, and the company was completely out of business by 1966.

Cincinnati

Helped considerably by its large German population, Cincinnati has always been a major brewing center. In fact, the "Queen City of the Ohio" might well have become this country's largest brewing city had it not been for Milwaukee's one very important and early advantage—a much greater supply of natural ice. Cincinnati, being fairly far south, simply did not have a very dependable supply of this all important commodity. As the following figures indicate, however, Cincinnati brewers have been considerable: 1879—18, 1890—24, 1910—22, 1935—11, 1950—5, 1960—3, and 1973—2. The two breweries continuing to keep alive Cincinnati's grand brewing tradition are the Hudepohl Brewing Co. and the Schoenling Brewing Co.

Two of the bottles pictured here are unusual items. The one on the left confirms the fact that Christian Moerlein had dreams of being a national brewer. This was an Albany, N. Y. distributor's bottle. The Hutchinson Jung bottle shows that some breweries were involved in related businesses long before the time of Prohibition.

Christian Moerlein was truly a great among brewers. Born in Truppach, Bavaria, in 1818, he emigrated to the United States in 1841. A year later he settled in Cincinnati and there operated a blacksmith shop for many years. Brewing was, however, his true love, for in 1853 he started a small brewery in partnership with Adam Dillmann. When Dillmann died in 1854, Moerlein teamed up with Conrad Windisch to form Moerlein and Windisch. Twelve years later Windisch left to co-found the equally famous Lion Brewery (see page 139) and Moerlein was on his own. And he did very well; by 1877 he was the thirteenth largest brewer in America, well ahead of such later giants as Anheuser-Busch, F. & M. Schaefer and Val Blatz. Christian Moerlein died in 1897, at the age of 79, and the company that bore his name eventually closed in 1919, at the age of 66.

The Jung Brewing Co., at 2011 Freeman Ave., traces its history back to 1857. It remained in operation until Prohibition.

The company that became Gerke Brewing Co. was founded in 1854, and was yet another Eagle Brewery in its early days. John Gerke became involved with the company in 1866 and, although he died in 1876, his heirs purchased the firm in 1881, changing the name (in 1882) to the Gerke Brewing Co. One of their brand names was "Social Session," a rather unusual name for a beer.

The John Kaufman Brewing Co. was located at 1622 Vine Street in the heart of Cincinnati. Its roots go back to 1856, and it remained in operation until Prohibition.

Last, but certainly not least, was John Hauck, a major Cincinnati brewery for over 50 years, from 1863 to Prohibition. John Hauck, himself, was the nephew of another famous Cincinnati brewer, George Herancourt. Exhibiting at fairs and expositions was a common practice among pre-Prohibition brewers, and John Hauck was no exception. To see this beautiful 1897 souvenir in color, turn to the color pages.

OHIO'S MODEL BREWERY
1897
Nashville Exposition
Souvenir,
The John Hauck Brewing Co.
Cincinnati, Ohio.
MAYER SCHREIBER
MOERLEINS' CINN
LAGER
ALBANY, N.Y.
THE JUNG BRG. CO.
MINERAL WATER
DEPT
CINCINNATI, O.
THE GERKE BREWING CO.
THE
CINCINNATI, O. U.S.A.
THE JOHN KAUFFMAN
BREWING CO.
CINCINNATI, O.

Detroit

While better known for automobiles than beer, Detroit has had an impressive number of breweries through the years. In 1879—30, 1890—33, 1910—19, 1935—15, 1950—6, 1960—6 and 1973—2.

Philip Kling, founded in 1856 and therefore one of Detroit's pioneer breweries, is represented here by a tray and embossed bottle. At one time various sections of the State of Michigan must have suffered from water problems. There are only two bottles in our large collection that stress "Pure and Without Drugs or Poisons"—and both are from Michigan. The Kling example is pictured here, and the second example is shown with the "Health Articles." Kling also operated after Repeal, from 1936 to 1946.

Altes was a brand name made famous in and around Detroit by the Tivoli Brewing Company, founded in 1897 by Franz Brogniez, a Belgian who had emigrated to the United States only one year earlier. Tivoli was a strong post-Prohibition brewing operation as well, remaining in business until 1948. Altes has since been brewed by the Altes Brewing Co., and National's Detroit Brewery, which still carries on the brand name.

While the label says 1889, published records give 1890 as the year Conrad Pfeiffer opened his brewery at 912 Beaufait Avenue. In any case, Pfeiffer, as with Tivoli, came back strongly after Prohibition, and brewed "Pfeiffer's

Famous Beer" until 1962, when the company merged with the E & B Brewing Co. (also of Detroit) to form the beginning of the Associated Brewing Co. complex.

Boy brewmaster, sailor, successful brewer—E. W. Voight was all of these. E. W. (his complete name is never given) grew up in his father's ale brewery at Madison, Wis. William Voight operated the Capitol Brewery there from 1854 to 1863. From 1860, when he was 16, to 1863, when his father sold the brewery, E. W. was the brewmaster. Then he set off for California to try his hand at being a sailor. In the fall of 1864 he returned from the coast to become a sailor on his father's Great Lakes vessel, a job held until 1866 when the elder Voight sold his shipping interest and returned to brewing, this time in Detroit. E. W. again joined the new company and took over control of the firm in 1871 when William Voight retired. Under E. W.'s leadership, the company grew dramatically and by 1902 it was a sizeable 100,000-barrel-capacity brewery. Unlike Tivoli and Pfeiffer, Voight (reopened as Voight-Prost) did not do well after Prohibition, lasting but four years, 1934-1937.

Detroit's remaining two breweries are the Stroh Brewing Co. and National Brewing Company's Detroit branch.

PREMISES OF THE VOIGT BREWERY COMPANY, LIMITED, DETROIT, MICHIGAN.

Milwaukee

Milwaukee—the very name conjures up visions of a foamy, bubbling cold glass of beer. Yet, despite the reputation, comparatively few breweries have made Milwaukee famous. In 1879 they numbered only 16, and in following years the figures decreased: 1890—14, 1910—12, 1935—11, 1950—6, 1960—6 and 1973—3. Yet what Milwaukee has lacked in numbers of brewers, it has made up for in quantity (and quality) of output. "The Cream City's" remaining breweries—Pabst, Schlitz and Miller—produce enough beer to make any city famous!

These rather interesting brewery shots are reprinted from an early 1900's booklet entitled "Prospectus of Hantke's Brewery School and Laboratories, at Milwaukee, the Brewing Center of the World." Hantke's operated in Milwaukee from 1898 to 1918, first at 646 Broadway and later at 200-210 Pleasant St. Conspicuous by its absence is Gettelman—probably because this brewery did not cooperate with the school in its programs.

Milwaukee-Waukesha's slogan, "The Imperial Health Beer," derives from the fact that the water used by the brewery came from the Waukesha Im-

FRED. MILLER
Brewing Co.

„Miller,
The Best Milwaukee Beer"

JUNG
BREWING CO.

„Jung Beer
Serves You
Right"

John Graf,

Weissbeer and
Soda Water.

"The Best What
Gives."

perial Spring, a health water source located on the property of the plant.

Blatz, once a nationally known and distributed beer, has had a choppy history during the past 15 years. Blatz, itself, sold out to Pabst in 1958. The United States government, however, then brought an anti-monopoly suit against Pabst, an action that was in and out of court for 11 years until Pabst was eventually forced to divest itself of Blatz. This it did in September of 1969 when the brand name was bought by the G. Heileman Brewing Co. of La Crosse, Wis. Heileman's goal is to build Blatz sales back to their former very substantial level.

Though the Jung Brewing Co. started relatively late, in 1896, their brewery plant had originally been the home of an earlier successful Milwaukee brewery—J. Obermann & Co. In fact, the plant, at 502 Cherry Street, was originally built in 1854. Philipp Jung, the founder, had formerly (1873-79) been brewmaster at Best & Co. (Pabst).

"The Best What Gives"—what a great slogan. And Graf actively used it, as witness the bottle illustrated in the "Bottles, A to Z" pages.

Milwaukee-Waukesha Brewing Co.

"The Imperial Health Beer."

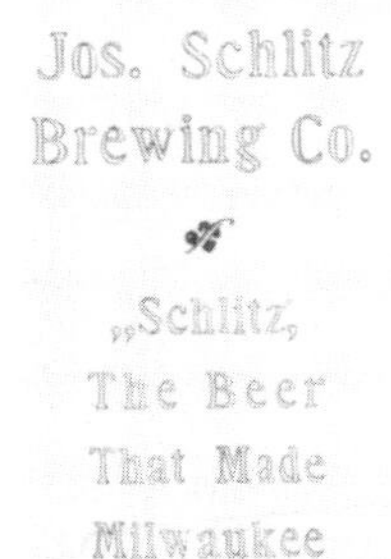

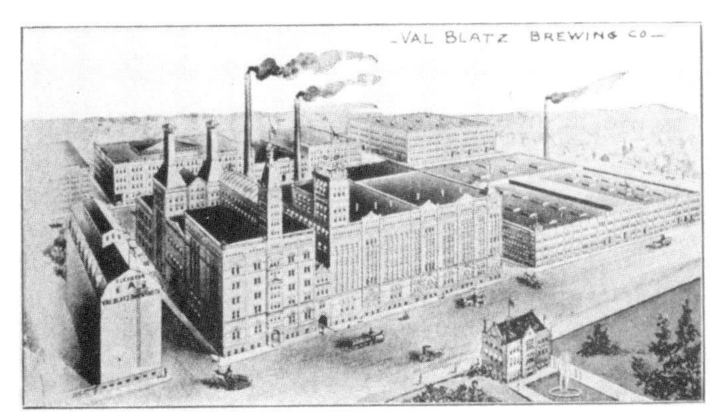

VAL. BLATZ Brewing Co.

„Blatz, The Star Milwaukee Beer"

New York City

New York has had numerous breweries over the years, but is now down to absolutely none (not including Brooklyn). The city was quite slow in getting started as a major brewing center because of the poor quality of its water. Not until the early 1840's, when the city began receiving water from upstate New York, did New York brewers really start to become a dominant force in Eastern brewing. Prior to that time most of New York City's beer came from the breweries of upriver towns—Peekskill, Poughkeepsie, Kingston, Newburgh, Albany and Troy. From the late 1870's to the present, however, the story has been one of constant decline: 1879—78 breweries, 1890—77, 1910—39, 1935—13, 1950—5, 1960—3, and 1973—0.

It is interesting to note that brewer Henry Elias (represented in the bottom row, third from the right) founded in the early 1870's the company that later became George Ringler & Sons Brewery (bottom row, second from left). The Henry Elias Brewing Co., itself, was not founded until 1874.

The Henry Haffen bottle (fourth from left, middle row) is a puzzle. Although records do not support it, Henry had to be associated with J. (John) and M. (Mathias) Haffen (second from right, middle row) in their Bronx brewery. Haffen is not a very common name and their addresses on East 152nd Street are virtually the same. Morrisania is, incidentally, the old name for that section of the Bronx.

The Beadleston & Woerz Empire Brewery (bottom row, fourth from right) is interesting in two respects. First, it was founded in 1845 as the branch of the Nash, Beadleston & Co. Brewery of Troy, N. Y. It was not until 1856 that the New York City brewery began to operate on its own. Second, it was built on the grounds of the old State Prison on West 10th Street and, in fact, a number of the former prison buildings were worked into the brewery complex as it was being built.

A. HUPFEL'S SON'S
161ST ST. & 3RD AVE.
NEW YORK
HENRY VON HINKEN & BRO
LAGER BEER
N.Y.
DAVID STEVENSON BREWING CO.
39TH TO 40TH ST.
10TH AVE.
NEW YORK.
V. LOEWER'S CAMBRINUS BREWERY CO.
A. KOCH
455
1ST AVENUE
NEW YORK
WALLACE BREWERY
400 CHERRY ST. N.Y.
DAVID MAYER
BREWING CO.
BOTTLING DEP'T
FULTON AVE.
JOHN FROHLICH
25
GREENWICH ST
N.Y.
PREMIUM
LAGER BEER
THE
EBLING
BREWING CO.
NEW YORK
U.S.A.
HENRY HAFFEN
MORRISANIA
NEW YORK
KOSTER & BIAL
CABINET
LAGER BEER
NEW YORK
J. & M. HAFFEN
NEW YORK
PREMIUM
LAGER BEER
HOWARD & CHILDS
BREWERS
NEW YORK CITY
BREWERS
NEW YORK
N.Y.
207 E. 54TH ST
NEW YORK
EMPIRE BREWERY
NEW YORK
NEW YORK
LAGER BEER
LAGER BEER
N.Y.
The John
Eichler
NEW YORK

Philadelphia

As can be seen by comparing the figures on numbers of breweries through the years, Philadelphia was the pre-Prohibition leader—1879—94, 1890—91, 1910—46, 1935—15, 1950—8, 1960—4, and 1973—2.

The Continental Brewing Co. (bottom row, extreme left, and top row, fourth from right) was the first brewery in the world—in 1879—to be lighted by electric power.

Charles Wolters founded, in 1877, the Prospect Brewing Co. (top row, extreme left); hence his name is included on the earlier blob Prospect bottle (top row, third from left). Prospect remained in operation until Prohibition.

Women's liberation was practiced in the brewing industry. The "E." in E. Vollmer (bottom row, second from left) stands for Elizabeth. She was either the wife or sister of August Vollmer, owner and operator of the brewery on Randolph Street from 1883 until his death in 1889. Elizabeth took over that year and guided the brewery successfully for many years thereafter, thus proving that brewing was not purely a man's world.

There are two effective examples of the use of brand name psychology on this page of illustrations. First is Philipp Hildenbrand's White Bear Brewery (bottom row, third from left). Exactly what a white bear has to do with beer is uncertain, but the image is certainly one of purity and cleanliness on the one hand, and strength and stamina on the other. Better yet, and more literal, is Welde & Thomas' Vacuum Purified Lager Beer (middle row, extreme right). It's doubtful whether even (John) Welde or (John) Thomas could have defined the process of vacuum purification, but it sounds impressive and undoubtedly considerably boosted sales.

Ortlieb's (tray on left of middle row) is the only brewery represented here that is still in business in 1973. A Philadelphia tradition, the plant on American Street goes back to the early 1860's, but it was in 1893 when Henry T. Ortlieb took over operation of the company that bears his name.

For information on John F. Betz and the Germania Brewing Co., see page 119.

In addition to Ortlieb's, Philadelphia's very proud brewing heritage is carried on by C. Schmidt & Sons. Schmidt is headquartered in Philadelphia and also operates breweries in Norristown, Pa., and Cleveland, Ohio.

THE PROSPECT BREWING CO PHILA.
J. SCHMIDTGALL 1826 NORTH 25TH ST. PHILADELPHIA
PROSPECT BREWERY CHAS. WOLTERS PHILADELPHIA
BREWING CO. N.W. COR. & THOMPSON STS. PHILADA, PA.
THE LOUIS BERGDOLL BREWING CO PHILA. PA.
Continental Brewing Company Philada
GERMANIA BREWING CO PHILADELPHIA
John F. Betz & Son Lim. Philada Pa.
J.F. BETZ & SON LIM. PHILA'DA

BREWERY
PHILADA. PA.
PREMIUM LAGER BEER
Ortlieb's
ALE
WELL BREWED AND AGED
HENRY F. ORTLIEB BREWING COMPANY
PHILADELPHIA
ORTLIEB'S BEER
XXX
Famous Manayunk
BEER
ALE
Liebert and Obert
BREWERS OF FINE BEER
SINCE 1873
MANAYUNK · PHILADELPHIA · PA.
LAGER BEER

TRADE MARK
PHILADELPHIA
E. VOLLMER
BREWERY
RANDOLPH ST.
PHIL'AD'A., PA.
WHITE BEAR BREWERY
16TH & ERIE AVE
PHILADA.
Poth's Beer
THEIS BREWERY
PHILADA.
Philada
J. STRAUBMULLER & SON PHILADA
J. STRAUBMULLER & SON PHILADA
John F. Betz
GERMANIA BREWERY
PHILADA.

St. Louis

While best known for two still-in-operation giants of the brewing industry—Anheuser-Busch and Falstaff—St. Louis has been the home of many other breweries, both large and small, through the years. 1879—23 breweries, 1890—29, 1910—25, 1935—11, 1950—6, 1960—3, and 1973—2.

These plant photographs are all reprinted from a book published before the turn-of-the-century, *E. Jungenfeld & Co., Brewers' Architects and Engineers, St. Louis.* It includes seventeen different St. Louis breweries (including Anheuser-Busch) designed and erected by Jungenfeld.

As explained earlier, British investors became heavily involved in the American brewing industry in the late 1880's. In St. Louis and East St. Louis a British syndicate known as St. Louis Breweries, Ltd., purchased eighteen breweries in 1889. The American name for this conglomerate was the St. Louis Brewing Association.

Of the breweries pictured, seven were merged into the St. Louis Brewing Association. These were Anthony & Kuhn (closed in 1899); Excelsior; Green Tree; H. Grone; Heim; Liberty (also closed about 1899); and Wainwright. The Green Tree Brewery took its unusual name from an adjoining building, The Green Tree Hotel.

Home Brewing Co., located at the junction of Miami and Salena Streets, was not involved in the syndicate's activities, being founded after the consolidation, in 1892. It was not a successful brewery, however, having closed between 1904 and 1909.

San Francisco

As the center of many commercial and industrial activities on the West Coast, San Francisco also has been the Far West's brewing capital. The years have taken their toll: 1879—38, 1890—26, 1910—19, 1935—4, 1950—6, 1960—6—but the city still boasts Anchor Steam Beer, General Brewing Co., and branches of Theodore Hamm and Falstaff.

Almost a century after this ad was written the wording sounds unbelievably formal and stuffy. Notice especially the phrase, "The amount of beer sold during the year 1881 was *about* 59,182 bbls." They counted right down to the last barrel, but still inserted "about."

How did the Philadelphia Brewery end up 3,000 miles away in San Francisco? The wanderings of brewer John Wieland provide the answer. He emigrated to the United States from Wurttemberg, Germany, in 1849. After a brief period of time in New York, he spent a year and a half in Philadelphia learning the bakery trade and, apparently, developing a real passion for the "City of Brotherly Love." In spite of this affection, he set out for the greener pastures of California in 1851. Wieland spent some time in the boom mining town of Nevada City before settling in San Francisco, where he once again entered the baking business. In August of 1855 he decided to try his hand at brewing, too, and purchased a part interest in the August Hoelscher & Co. Brewery. A year later he started his own company on Second Avenue, naming it the Philadelphia Brewery.

Wieland proved to be a good businessman and brewer, and the company rapidly became the largest brewery in California and, in fact, on the entire West Coast. By 1879, a year for which sales figures are available, Wieland sold 44,276 barrels, more than double the amount of the second largest Coast brewery—H. Aherns' Chicago Brewery (also located in San Francisco), which amassed the figure of 20,261 barrels sold. Ironically, another large San Francisco brewery (with an even 13,000 barrels sold in 1879) was F. Hagerman's Albany Brewery. Although the rest of the nation seemed to be well represented, there was at the time no San Francisco brewery in San Francisco. The same situation prevailed in Los Angeles where there were but three breweries—named New York, Philadelphia, and United States.

John Wieland died in January of 1885, and his three sons took over operation of the business. Five years later, they sold out (along with nine other California breweries) to an English syndicate, San Francisco Breweries, Ltd. The British-controlled Wieland's apparently never officially re-opened after Prohibition. The firm is listed in the 1934 *Brewers Journal Directory*, but was never able to get a Federal Permit. Thus, for all intents and purposes, this colorful and very significant brewery was pre-Prohibition only.

472 *Representative Business Houses.*

PHILADELPHIA

BREWERY

Corner Second and Folsom Streets,

SAN FRANCISCO.

The Amount of Beer sold during the year 1881 was about 59,182 bbls.

I take the present opportunity of thanking my Friends and Customers for the liberal support heretofore extended to the

PHILADELPHIA BREWERY,

And notify them that I have added to my establishment

New and Extensive Buildings,

By which I hope, through the greatly increased facilities now possessed by me, to furnish, as usual, a

Superior Article of Lager Beer

That shall not only equal that previously furnished by me but convince them that I am determined to merit their continued patronage and support.

JOHN WIELAND.

Interesting Facts About Beer & Brewing

The word "beer" comes from the Latin "bibere," meaning "to drink." In numerous other countries the base is the same but the spelling somewhat different:

France and Belgium–Biere	Italy–Bierra
Germany and Holland–Bier	Japan–Biru
Indonesia–Bir	Israel–Bior

Many of our famous forebears were brewers, either commercially or for their own personal use. William Penn had his own private brewery at Pennsbury, his country estate on the Delaware River. Samuel Adams, "Father of the Revolution," owned and operated a brewery that he had inherited from his father. Before he gained fame as a Revolutionary War general, Israel Putnam was a Connecticut brewer. And the most famed of all Revolutionary War generals and "Father of our Country," George Washington, had a personal recipe for the beer that was brewed at Mount Vernon, his Virginia estate.

If it hadn't been for beer, Plymouth Rock would probably be just another oversized and uncared for pebble in the ocean. It is believed that, instead of continuing farther south as originally planned, the Pilgrims landed the Mayflower at Plymouth because they were running short of essential supplies, especially beer. One of those who arrived in the New World on the Mayflower was John Alden. Alden, of the fabled Priscilla Mullins/Miles Standish "Speak for yourself, John" story, had come along as cooper to care for the beer barrels.

Approximately 50 percent of the total United States adult population can be classified as beer drinkers. Of the beer universe, 63 percent are men and 37 percent are women. On an average day, almost 13 percent of the total adult population has a beer. As might be expected, this percentage increases during the summer (14.6 percent) and decreases during the winter (10.6 percent).

The better to transport his beer from the brewery to neighboring taverns, a Dutch brewer in New Amsterdam paved the road leading from his brewhouse with cobblestones in 1630. Called Stone Street, this was America's first paved road.

Two of today's brewing giants, Pabst and Miller, were founded by the same brewing family. In 1844 Jacob Best, Sr., formed what has since become the Pabst Brewing Company. Six years later, in 1850, two of his sons, Charles and Lorenz Best, founded the Menomonee Valley Brewery. Also known as the Plank-Road Brewery, this firm was sold in 1855 to Frederick Miller, who had just emigrated to Milwaukee from Germany. We know this firm today, of course, as the Miller Brewing Company.

Believe it or not, forty-six of the then forty-eight states ratified the Eighteenth Amendment (Prohibition). Mississippi was the first, on January 8, 1918, and New Jersey was the last, on March 9, 1922 (over two years after Prohibition had already gone into effect). Surprisingly, the only two states that did *not* ratify what must be considered our country's most unfortunate amendment were the generally rather conservative New England states of Connecticut and Rhode Island.

Famed movie and television actress Shirley Jones is a brewer's daughter. Shirley's father, Paul Jones, was President of the Jones Brewing Company of Smithton, Pennsylvania, from 1952 until his death in 1959.

The United States Brewers Association is the oldest still-in-existence trade organization in the United States. It was founded well over one hundred years ago, in 1860, to set up a system of beer taxation that was fair for both the federal government and the brewers.

The word "bridal" is a derivation of "bride ale," which was the name for old-time English wedding feasts at which ale, naturally, was the featured beverage.

Another expression rooted in English beer drinking tradition that has found its way into our language is "Watch your P's and Q's." In days of old, tavern keepers would mark each purchase on the wall or a board. Purchases were generally by the pint (P's) or quart (Q's)—and barmaids (and customers) were always reminded to "Watch your P's and Q's"; i.e., keep track of how much money was owed.

Everyone knows that Francis Scott Key wrote the *words* to "The Star Spangled Banner." But did you know that he used an old drinking song for the *tune?* Called "To Anacreon In Heaven," this was the club song for a high-spirited, early 1800's London drinking club known as the Anacreontic Society.

Vassar College, America's first privately endowed college for women, was founded by a Poughkeepsie, N. Y. ale brewer, Matthew Vassar. Matthew's father, James, had started brewing in the late 1790's. From 1810 on Matthew was a partner in the company, although he began to lose interest in brewing in the 1850's as he pursued the idea of an institution of higher learning for women. In 1861 his dream came true—Vassar Female College was chartered. Six years later, in 1867, the name was changed to its present-day Vassar College.

While the college Vassar endowed has fared well through the years, his brewery did not. Refusing to switch to the ever more popular lager, the brewery limited its output to beer in the British tradition—ale, porter and stout. This proved disastrous, and the once prosperous Vassar Brewery closed its doors in 1899, thirty-one years after Matthew Vassar's death in 1868. He still lives on, however, in a Vassar song:

And so you see, for old V. C.
Our love shall never fail.
Full well we know
That all we owe
to Matthew Vassar's ale!

Peerless was a name long respected in the automotive world. For many years, until 1931, it produced fine cars at its plant in Cleveland, Ohio. In 1931 automobile production stopped and the plant lay idle for two years until September of 1933, when Clevelanders learned of an interesting conversion. The Brewing Corporation of America was taking over the old Peerless plant and converting it into a brewery. And that they did. Better known now as Carling (the company's official name since 1954), the company brewed in Cleveland until the early 1970's. At present the plant is a branch brewery of C. Schmidt and Sons. Fine cars and fine beer—two of the pleasures of life and both produced in the same plant!

Here's to a long life and a merry one,
A quick death and easy one
A pretty girl and a true one
A cold beer—and another one.

So goes an old toast that pretty much sums up the spirit of *The Beer Book.*

Supposedly the term "toast" comes from the fact that, in days of old, beer was often consumed in front of the fireplace, where bread was being toasted at the same time. To add nutrition and flavor (?) bits of the toast would be thrown in the about-to-be-drunk beer—and then a "toast" would be made. In any case the custom of toasting one's drinking companion goes back many centuries. Here are two toasts from a booklet published by the William J. Lemp Brewing Company in 1896.

LET'S LIVE.

While we live, let's live in clover,
For when we're dead, we're dead all over.

WOMEN AS A TOAST.

Drink to fair woman, who, I think,
Is most entitled to it;
For if anything drives men to drink,
She certainly can do it.

LIST OF OPERATING BREWERIES BY STATES

While the trend continues to be toward fewer and fewer (but larger and larger) breweries, there were still 122 plants licensed to operate in June of 1973. Several of these plants have already closed, and several more undoubtedly will before the year ends. This is the official United States Department of the Treasury listing.

ARIZONA

National Brewing Co., The
150 South Twelfth St.
Phoenix

CALIFORNIA

Anchor Steam Beer
541 8th St.
San Francisco

Anheuser-Busch, Inc.
15800 Roscoe Blvd.
Los Angeles

Falstaff Brewing Corporation
470 Tenth St.
San Francisco

Falstaff Brewing Corporation
1025 West Julian St.
San Jose

General Brewing Company
500 E. Commercial St.
Los Angeles

Hamm Company, Theodore
1550 Bryant St.
San Francisco

General Brewing Company
2601 Newhall St.
San Francisco

Miller Brewing Company
819 N. Vernon Ave.
Azusa

Pabst Brewing Company
1910-2026 N. Main St.
Los Angeles

Schlitz Brewing Company, Jos.
7521 Woodman Ave.
Los Angeles
Van Nuys P. O.

COLORADO

Coors Company, Adolph
Golden

Walter Brewing Company, The
Hickory & LaCrosse Sts.
Pueblo

CONNECTICUT

Hull Brewing Company, The
820 Congress Ave.
New Haven

FLORIDA

Anheuser-Busch, Inc.
111 Busch Drive
Jacksonville

Anheuser-Busch, Inc.
3000 August A. Busch, Jr. Blvd.
Tampa

Duncan Brewing Company, Inc.
202 Gandy Road
Auburndale

National Brewing Co., The
637 N. W. 13th St.
Miami

Old Munich Brewing Company
16500 N. W. 52nd Ave.
Hialeah

Schlitz Brewing Company, Jos.
11111 30th St.
Tampa

GEORGIA

Carling Brewing Company Incorporated
3599 Browns Mill Road SW
Atlanta

Pabst Brewing Company
Pabst

HAWAII

Honolulu Sake Brewery & Ice Co., Ltd.
2150 Booth Road
Honolulu

Schlitz Brewing Company, Jos.
98051 Kamehameha Highway
Aiea, Oahu

ILLINOIS

Carling Brewing Company
1201-25 West "E" St.
Belleville

Peter Hand Brewing Co.
1000 West North Ave.
Chicago

Pabst Brewing Company
4541 Prospect Road
Peoria Heights

INDIANA

Falstaff Brewing Corporation
1019-1051 Grant Ave.
Fort Wayne

Heileman Brewing Co. of Indiana, Inc., G.
1301 W. Pennsylvania St.
Evansville

Old Crown Brewing Corporation
2501-38 Spy Run Ave.
Fort Wayne

IOWA

Dubuque Star Brewing Co.
East 4th St. Extension
Dubuque

KENTUCKY

Falls City Brewing Company
3050 W. Broadway
Louisville

Wiedemann Brewing Company, The Geo., Div. of G. Heileman Brewing Company, Inc.
601 Columbia St.
Newport

LOUISIANA

Dixie Brewing Company, Inc.
2401 Tulane Ave.
New Orleans

Falstaff Brewing Corporation
2600 Gravier St.
New Orleans

Jackson Brewing Co.
620 Decatur St.
New Orleans

MARYLAND

Carling Brewing Company
Baltimore Beltway at Hammond's Ferry Road
Baltimore

National Brewing Company, The
3602 O'Donnell St.
Baltimore

Queen City Brewing Company, The
208 Market St.
Cumberland

Schaefer Brewing Co., The F. & M.
1101 S. Conkling St.
Baltimore

MASSACHUSETTS

Carling Brewing Company
1143 Worcester St.
Natick

Piel Bros. Inc.
45-95 North Chicopee St.
Willimansett

Jacob Ruppert, Inc.
29-43 Brook St.
New Bedford

MICHIGAN

Bosch Brewing Company
Adams Township
Houghton, P. O.

Carling Brewing Company
907 S. Main St.
Frankenmuth

Geyer Bros. Brewing Co.
415 Main St.
Frankenmuth

National Brewing Co., The
3765 Hurlburt Ave.
Detroit

Stroh Brewing Company, The
909 E. Elizabeth St.
Detroit

MINNESOTA

Cold Spring Brewing Company
219 North Red River St.
Cold Spring

Grain Belt Breweries, Inc.
1215 Marshall St., N.E.
Minneapolis

Hamm Company, Theodore
720 Payne Ave.
St. Paul

Heileman Brewing Co., Inc., G.
882 W. 7th St.
St. Paul

Schell Brewing Co., August
South Payne St.
New Ulm

MISSOURI

Anheuser-Busch, Inc.
721 Pestalozzi St.
St. Louis

Falstaff Brewing Corporation
1920 Shenandoah Ave.
St. Louis

Pearl Brewing Company
603 Albermarle St.
St. Joseph

Schlitz Brewing Company, Jos.
316 Oak St.
Kansas City

NEBRASKA

Falstaff Brewing Corporation
25th St. and Deer Park Blvd.
Omaha

NEW HAMPSHIRE

Anheuser-Busch, Inc.
1000 Daniel Webster
Highway
Merrimack

NEW JERSEY

Anheuser-Busch, Inc.
200 U. S. Highway 1
Newark

Champale, Inc.
Lalor & Lamberton Sts.
Trenton

Eastern Brewing
Corporation
329 N. Washington St.
Hammonton

Pabst Brewing Company
391-399 Grove St.
Newark

Rheingold Breweries, Inc.
119 Hill St.
Orange

NEW YORK

Genesee Brewing Co.,
Inc., The
100 National St.
Rochester

Koch Brewery, Fred
15-25 W. Courtney St.
Dunkirk

Piel Bros. Inc.
315 Liberty Ave.
Brooklyn

Rheingold Breweries, Inc.
36 Forrest St.
Brooklyn

Schaefer Brewing Co.,
The F. & M.
430 Kent Ave.
Brooklyn

Schaefer Brewing Co.,
The F. & M.
24 to 28, 34 and 51 North
Ferry St.
Albany

West End Brewing Com-
pany of Utica, N. Y., The
811 Edward St.
Utica

NORTH CAROLINA

Schlitz Brewing Company,
Jos.
4791 Schlitz Ave.
Winston-Salem

OHIO

Anheuser-Busch, Inc.
700 East Schrock Road
Columbus

Hudepohl Brewing
Company, The
Fifth & Gest Sts.
Cincinnati

Schmidt & Sons, Inc., C.
9400 Quincy Ave.
Cleveland

Schoenling Brewing
Company, The
1625 Central Parkway
Cincinnati

Wagner Breweries, Inc.,
August
605-631 S. Front St.
Columbus

OREGON

Blitz-Weinhard Company
1133 W. Burnside St.
Portland

PENNSYLVANIA

Altoona Brewing
Company
1718-32 N. Ninth Ave.
Altoona

Du Bois Brewing
Company, The
Hahne's Court, S. Main St.
Du Bois

Erie Brewing Company,
The
2124-2212 State St.
Erie

Fuhrmann & Schmidt
Brewing Co.
235-249 S. Harrison St.
Shamokin

Horlacher Brewing
Company
311 Gordon St.
Allentown

Jones Brewing Company
Second St. & B&O R.R.
Smithton

Latrobe Brewing Company
119 Ligonier St.
Latrobe

Lion, Inc., The
5 and 6 Hart St.
Wilkes-Barre

Mount Carbon Brewery
716 Centre St.
Borough of Mount Carbon
Pottsville, P. O.

Ortlieb Brewing Company,
Henry F.
824 N. American St.
Philadelphia

Pittsburgh Brewing
Company
3340 Liberty Ave.
Pittsburgh

Reading Brewing
Company
S. 9th & Little Laurel Sts.
Reading

Schaefer Brewing Co.,
The F. & M.
S/W Corner of Rte. 22 &
Highway 100
Allentown

Schmidt & Sons,
Incorporated, C.
127 Edward St.
Philadelphia

Schmidt & Sons,
Incorporated, C.
151 W. Marshall St.
Norristown

Stegmaier Brewing
Company
Market & Baltimore Sts.
Wilkes-Barre

Straub Brewery, Inc.
Rear 303 Sorg St.
St. Marys

Yuengling & Son,
Incorporated, D. G.
5th & Mahantongo Sts.
Pottsville

RHODE ISLAND

Falstaff Brewing
Corporation
Garfield Ave. &
Cranston St.
Cranston

TENNESSEE

Schlitz Brewing Company,
Jos.
5151 Raines Rd.
Memphis

TEXAS

Anheuser-Busch, Inc.
775 Gellhorn Drive
Houston

Falstaff Brewing
Corporation
3301 Church St.
Galveston

Lone Star Brewing
Company
542 Lone Star Blvd.
San Antonio

Miller Brewing Company
7001 South Freeway
Fort Worth

Pearl Brewing Company
312 Pearl Parkway
San Antonio

Schlitz Brewing Company,
Jos.
1400 West Cotton St.
Longview

Spoetzl Brewery, Inc.
Brewery St., Block 60
East End Addition
Shiner

VIRGINIA

Anheuser-Busch, Inc.
Williamsburg

Champale Products Corp.,
The
710 Washington Ave.
Norfolk

WASHINGTON

Carling Brewing Company,
Incorporated
2120-42 South C. St.
Tacoma

General Brewing Company
615 Columbia St.
Vancouver

Olympia Brewing
Company
Tumwater, P.O. Box 947
Olympia

Rainier Brewing Company
3100 Airport Way
Seattle

WISCONSIN

Heileman Brewing Co.,
Inc., G.
1000-1028 S. Third St.
La Crosse

Heileman Brewing Co.,
Inc., G.
1012 New York Ave.
Sheboygan

Huber Brewing Co., Jos.
1200-1208 14th Ave.
Monroe

Leinenkugel Brewing Co.,
Jacob
1-3 Jefferson Ave.
Chippewa Falls

Miller Brewing Company
4002-4026 W. State St.
Milwaukee

Pabst Brewing Company
917 W. Juneau Ave.
Milwaukee

Peoples Brewing
Company, The
1506-12 S. Main St.
Oshkosh

Rice Lake Brewing Co.
816 Hammond Ave.
Rice Lake

Schlitz Brewing Company,
Jos.
235 W. Galena St.
Milwaukee

Stevens Point Brewery
2617 Water St.
Stevens Point

Walter Brewing Company
318 Elm St.
Eau Claire

Bibliography

Books

Anderson, Sonja and Will. *Anderson's Turn-of-the-Century Brewery Directory.* Carmel, N. Y., 1968.

Anderson, Sonja and Will. *Beers, Breweries and Breweriana.* Carmel, N. Y., 1969.

Barley, Hops and History. New York: United States Brewers Foundation, n. d.

Baron, Stanley. *Brewed in America.* Boston: Little, Brown & Co., 1962.

Birmingham, Frederic. *Falstaff's Complete Beer Book.* New York: Award Books, 1970.

Breweries of Otto C. Wolf. Philadelphia, 1906.

Cochran, Thomas C. *The Pabst Brewing Company.* New York: New York University Press, 1948.

Ehret, George. *Twenty-five Years of Brewing.* New York, 1891.

A History of Packaged Beer and Its Market in the United States. New York: American Can Company, 1969.

Kelley, William J. *Brewing in Maryland.* Baltimore, 1965.

Kroll, Wayne L. *Wisconsin Breweries and Their Bottles.* Jefferson, Wis., 1972.

Munsey, Cecil. *The Illustrated Guide to the Collectibles of Coca-Cola.* New York: Hawthorn Books, 1972.

Munsey, Cecil. *The Illustrated Guide to Collecting Bottles.* New York: Hawthorn Books, 1970.

Muzio, Jack. *Collectible Tin Advertising Trays.* Santa Rosa, Calif., 1972.

Myers, Robert. *Some Thoughts on the Beer Can's First 35 Years.* Santa Barbara, Calif., 1971.

The New York City Record of the Wholesale and Retail Wine and Liquor Dealers, Ale and Beer Brewers, Mineral Water Dealers, Segar and Tobacco Manufacturers of the City and County of New York. New York: New York City Record Publishing Co., 1885.

One Hundred Years of Brewing. Chicago: H. S. Rich & Co., 1903.

Portfolio of Breweries and Kindred Plants Designed and Erected by E. Jungenfeld & Co. St. Louis, c. 1900.

Romer, Frank. *Reviewing American Brewing.* Baltimore: Crown Cork and Seal Co., 1942.

Salem, F. W. *Beer, Its History and Its Economic Value as a National Beverage.* Hartford: F. W. Salem & Co., 1880.

Souvenir of Philadelphia. Prepared for the Thirty-sixth Annual Convention of the United States Brewers Association. Philadelphia, 1896.

Weeks, Morris, Jr. *Beer and Brewing in America.* New York: United States Brewers Foundation, 1949.

Many brewers' directories, handbooks and bluebooks, 1882-1973.

Articles

"The A. Gettelman Brewing Company, a Century of Brewing." *Brewers Digest* (Sept., 1954).

"Brewing." *Encyclopedia Britannica* (Chicago, 1964).

Burck, Charles G. "While the Big Breweries Quaff, the Little Ones Thirst." *Fortune* (Nov., 1972).

"Fact Sheet–Bock Beer." *Michigan Beverage News* (Feb. 15, 1971).

Gaylord, Bill. "Glasses–Name Your Poison!" *The Western Collector* (May, 1968).

McBride, John R. "Trifles and Treasures for Drinking the Golden Brew." *Mankind* (Vol. II, No. 1).

North American Wine and Spirit Journal. Boston (April, 1909).

Patterson, Schuyler. "Back of the Bock Beer Tradition." *Brewer and Maltster* (Feb., 1937).

Schiller, Ronald. "A Heady History of Beer." *Playboy* (Sept., 1972).

Smith, Don. "Buffalo Hunting." *The Western Collector* (Mar., 1972).

"Steam Beer." *Modern Beverage Age* (Nov., 1960).

"The Story of Bock Beer." *Illinois Beverage Journal* (Mar. 1963).

"A Struggle to Stay First in Brewing." *Business Week* (Mar. 24, 1973).

"Trouble Brewing." *Newsweek* (July 23, 1973).

Uihlein, Robert A., Jr. "That Milwaukee Slogan Stems From Chicago's Fire." *Nation's Business* (Jan., 1970).

Newspaper articles

Delfiner, Rita. "Bottle and Can Dispute Rages Again." *The New York Post* (Feb. 16, 1972).

Federman, Stanley. "Returnables: Oregon Throw Away–Bottle Ban May Set National Pattern." *The National Observer* (Mar. 24, 1973).

Hamill, Pete. "Beer Crisis." *The New York Post (Feb. 21, 1973).*

Hillinger, Charles. "Only 1 Steam Beer Brewery Left, But It's Making Comeback." *The Boston Globe* (Mar. 10, 1972).

"Why Local Brewers Are In A Froth" and "Can the New York Brews Keep Their Heads?" Two-part series. *The New York Daily News* (Apr. 10, 11, 1973).

LIST OF TRAY MANUFACTURERS

Since many tray manufacturers included their name on the inside bottom rim of their product, knowing the dates of operation of the various companies is very helpful in dating trays. Listed here, with their dates of operation, are the major manufacturers of American trays. Much of the data was gathered from *Collectible Tin Advertising Trays* by Jack Muzio.

A. C. Co. (American Colortype Co.), Newark, N. J., 1930–?
American Art Works, Coshocton, Ohio, 1909–1950
American Can Co., Greenwich, Conn., 1901–still in business
Bachrach Co., San Francisco, Calif., 1895–1917?
H. D. Beach Co., Coshocton, Ohio, 1901–still in business
The Burdick Co., New York, N. Y., 1929–still in business, but ended tray manufacturing in the 1940's
Canadian-American Art Works, Ltd., Montreal, Quebec, Canada, 1912-1946
Electro-Chemical Engraving Co., New York, N. Y., 1900–still in business as the Superior Metal Lithography Corp.; no tray manufacturing since the early 1950's
Haeusermann Litho Co. (Haeusermann Metal Manufacturing Co.), New York, N. Y. and Chicago, Ill., 1905-1921
Mayer and Lavenson Co., New York, N. Y., 1903–1914
The Meek Co., Coshocton, Ohio, 1901–1909
The Meek and Beach Co., Coshocton, Ohio, 1901–1905
Chas. W. Schonk Co., Chicago, Ill., 1890–c. 1935
Sentenne and Green, New York, N. Y., 1894–1923
The Standard Advertising Co., Coshocton, Ohio, 1888–1901
The Tuscarora Advertising Co., Coshocton, Ohio, 1887-1901

Index

CP refers to the color pages